Mohammad M. Banat • Sara Paiva
Editors

Smart Technologies for Smart Cities

Springer

Editors
Mohammad M. Banat
Department of Electrical Engineering
Jordan University of Science
and Technology
Irbid, Jordan

Sara Paiva
Instituto Politécnico de Viana do Castelo
School of Technology and Management
Viana do Castelo, Portugal

ISSN 2522-8595 ISSN 2522-8609 (electronic)
EAI/Springer Innovations in Communication and Computing
ISBN 978-3-030-39985-6 ISBN 978-3-030-39986-3 (eBook)
https://doi.org/10.1007/978-3-030-39986-3

This Springer imprint is published by the registered company Springer Nature Switzerland AG.
The registered company address is: Gewerbestrasse 11, 6330 Cham, Switzerland

EAI/Springer Innovations in Communication and Computing

Editor's Note

The impact of information technologies is creating a new world yet not fully understood. The extent and speed of economic, life style and social changes already perceived in everyday life is hard to estimate without understanding the technological driving forces behind it. This series presents contributed volumes featuring the latest research and development in the various information engineering technologies that play a key role in this process.

The range of topics, focusing primarily on communications and computing engineering include, but are not limited to, wireless networks; mobile communication; design and learning; gaming; interaction; e-health and pervasive healthcare; energy management; smart grids; internet of things; cognitive radio networks; computation; cloud computing; ubiquitous connectivity, and in mode general smart living, smart cities, Internet of Things and more. The series publishes a combination of expanded papers selected from hosted and sponsored European Alliance for Innovation (EAI) conferences that present cutting edge, global research as well as provide new perspectives on traditional related engineering fields. This content, complemented with open calls for contribution of book titles and individual chapters, together maintain Springer's and EAI's high standards of academic excellence. The audience for the books consists of researchers, industry professionals, advanced level students as well as practitioners in related fields of activity include information and communication specialists, security experts, economists, urban planners, doctors, and in general representatives in all those walks of life affected ad contributing to the information revolution.

About EAI

EAI is a grassroots member organization initiated through cooperation between businesses, public, private and government organizations to address the global challenges of Europe's future competitiveness and link the European Research community with its counterparts around the globe. EAI reaches out to hundreds of thousands of individual subscribers on all continents and collaborates with an institutional member base including Fortune 500 companies, government organizations, and educational institutions, provide a free research and innovation platform.

Through its open free membership model EAI promotes a new research and innovation culture based on collaboration, connectivity and recognition of excellence by community.

More information about this series at http://www.springer.com/series/15427

Preface

Smart technologies currently play major roles in society and are the focus of several scientific proposals conducted by researchers. Those proposals can be applied to several domains, such as transport, mobility, culture, and health. Such proposals have been enabled by the evolution of electronic equipment and devices, as well as paradigms such as the Internet of Things, cloud, or edge computing. Smart technologies will allow carrying cities to a whole new level, giving birth to smart cities, which will be a reality in most countries in the next years to come.

In this book we've gathered valuable contributions to help the reader understand the latest research in the area of smart technologies and smart cities. In this context, this book includes two main parts: smart cities applications and smart mobility.

In Part I, chapter "Innovation Accelerator to Make Portuguese Parishes Smarter" presents a study on how smart technologies can be introduced in parishes and also includes the validation of the proposed framework in a Portuguese parish. Chapter "A Novel Intelligent WEB Application on Refrigerator Systems" focuses on smart homes and introduces a novel mini-smart refrigerator to be used in hotels or village farmhouses. Chapter "ChildPOPS: A Smart Child Pocket Monitoring and Protection System" focuses on a smart system to monitor advanced lifestyles of infants. Chapters "Signal Processing-Based Automated Fault Detection Methods for Smart Grids" and "Monitoring and State Estimation of Modern Power Systems" are both related to power systems and how they can become more reliable and contribute to better energy consumption in future cities.

Part II is related to a specific field of smart cities which is smart mobility. Chapter "A Vehicular Internet of Things (IoT) System for High-Granularity Air Quality Monitoring in Smart Cities" proposes an IoT for air quality monitoring that can be replicated on vehicles. Chapter "Renewable Mobility in Smart Cities: The MOVESMART Approach" proposes several cloud-based mobility services delivered through a mobile application that also offers a route planning service. Chapter "Intelligent System for Predicting Motorcycle Accident by Reaching into a Smart City Using a Kriging Model to Achieve Its Prevention and the Reduction of Deaths in the Medium Term" presents the projections for the next 10 years modeled based on the repository of traffic accidents related to Italian scooters.

Chapter “Development of a Java Library to Solve the School Bus Routing Problem” presents a Java library designed to solve the school bus routing problem. Finally, in chapter “An Efficient Hybrid Evolutionary Algorithm for the Smart Vehicle Routing Problem” authors investigate real-time routing of vehicles serving multiple clients and saving energy consumption.

The chapters in this book have been prepared by contributors from several countries. The editors would like to thank all authors for their contributions and all reviewers for their fine reviewing efforts. Special thanks are due to Ms. Eliška Vlčková, Managing Editor of European Alliance for Innovation (EAI), for all her support and cooperation throughout the process of creating this book.

Irbid, Jordan — Mohammad M. Banat
Viana do Castelo, Portugal — Sara Paiva

Contents

Part I
Smart Cities Applications

Innovation Accelerator to Make Portuguese Parishes Smarter

Inês Vizela, Emanuel Costa, and Vítor Santos

1 Introduction

During the last decades of the twentieth century, new concepts for public services arose, and associated with them also arose concepts like trust, collaboration, participation, and liability as founding stones in administration and citizens' relationships so there was the compulsory need to grant more power to public administration and to get the citizens more involved in all processes around them [1]. The World Wide Web concept was defined as becoming an essential and revolutionary part of citizens' daily lives in different contexts and places. A kind of paradigm shift had occurred which leads to changes not only in the dissemination but also in processing information [1]. Crowdsourcing is also an interesting topic to be evaluated in the context of this study as a way to use citizens and stakeholders' knowledge to solve municipal problems like the ones they experience during their daily routines. All of this, through e-participation. There was the need for some expertise on this topic inside parishes in order to gather as many innovative and creative contributions as possible [2]. A smart place had been considered the one that combines perfectly the physical and the virtual world in only one location, where information and communication technologies are considerably essential in order to make the bridge between these two realities [3]. All this new integration of technologies in citizens' lives derived into two new research areas related to e-governance, such as how technology mediates citizens' participation or e-participation [4]. But, most of the studies were much more focused on the website of these municipalities and not

I. Vizela (✉) · V. Santos
NOVA IMS, Lisbon, Portugal
e-mail: ines.vizela@outlook.com; vsantos@novaims.unl.pt

E. Costa
Área Metropolitana de Lisboa, Lisbon, Portugal

M. M. Banat, S. Paiva (eds.), *Smart Technologies for Smart Cities*,
EAI/Springer Innovations in Communication and Computing,
https://doi.org/10.1007/978-3-030-39986-3_1

in the overall internal processes and interaction with the citizens. Although this question has been addressed in main cities/municipalities but neglected at a local level, unfortunately, and despite the implementation of dozens of smart city projects, citizens are still far away from their parish councils [5], and many times, they are made aware of what is being implemented at their own city through the media and not because of a good communication with their parishes. Desires to the near future were also defined on how it should be and how citizens' proximity shall be valued; regarding that, the Portuguese Republic said in 2018 that public administration power should be decentralized from central to local administration [6] which may be understood as a step ahead on taking citizens and local administration closer (Fig. 1). This improvement will only be possible after taking advantage of "the relationship between municipalities and the government" because, nowadays, it "is very scarce" which distances local and central administration [7] which may have caused a lack of communication which directly affected citizens. So, the main question that will be addressed with this study is "How to make Portuguese parishes smarter?"; to achieve the answer, there was a need of an understanding on what kind of technologies, innovation techniques, and methods could be useful to obtain a smarter parish. The feeling of transparency between the two main actors of this

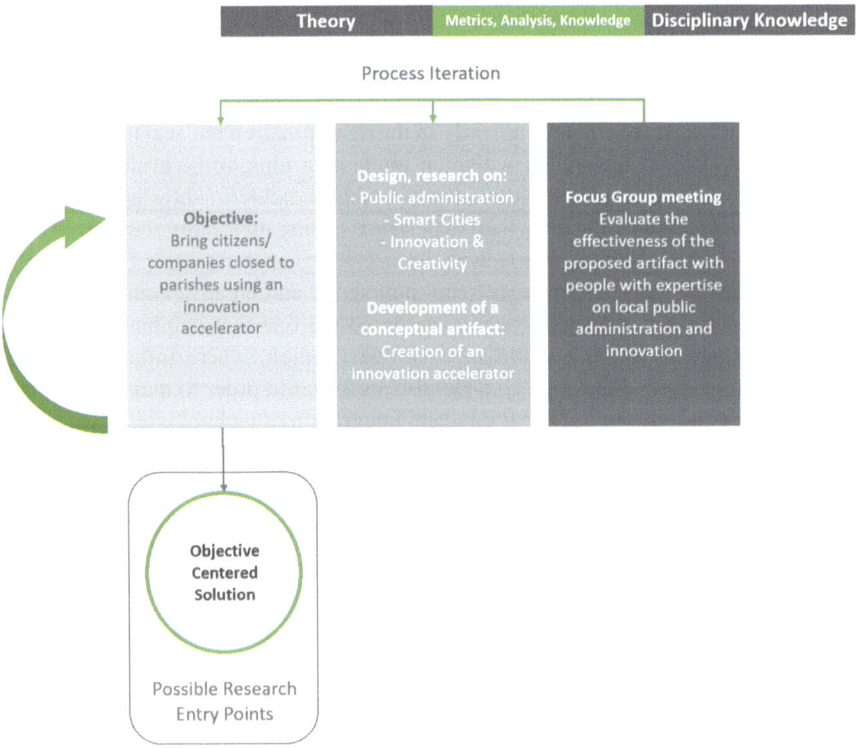

Fig. 1 DSR implementation strategy

study has been considered as a key element for the success of local e-government projects and also to be able to lead with a smart democracy system [8]. The reasons for conducting this research were the nonexistence of a smart concept for parishes and the fact that most of smart initiatives were implemented in the context of a municipality and not at a parish level. Parishes were also chosen as the focus of this study as they were considered the most important way to link citizens with public administration, even though this link was made, most of the time, through traditional clerk services. So, parishes needed to become part of today's information society in order to take some advantage from it [9]. An advantage for this study is the fact that the Portuguese smart cities section was considered the biggest one inside the National Association of Portuguese Municipalities [10], which leads us to think that Portugal invests in smartness and to hope that the output artifact of this study may be implemented in parishes in the future. As another motivation to start this study, there is the fact that e-governance in Portugal has been considered an unknown topic and the majority of parishes does not know how they are supposed to apply this kind of twenty-first-century hot topics to their daily routines [9]. Another motivational factor to go ahead with this study around innovation for Portuguese parishes was knowing that it was already part of "Lisbon agenda" the aim to create "a knowledge-based economy in Europe driven by innovation" [11]. This chapter is organized as follows: First, all theoretical framework that gave support to this study including public administration, smart cities, innovation, and creativity is exposed. Next, the methodology followed explaining how it applies to this study. Then, a model of an innovation accelerator for smart cities parishes that arose from the realization of this study is explained in detail as well as the discussion process required for the final model validation. Finally, the study's conclusions are presented as well as the future work with recommendations.

2 Theoretical Framework

To start introducing the problem behind this study, there was the need for a theoretical background on topics like public administration, smart cities, as well as innovation and creativity.

2.1 Public Administration

At the global level, there was a need to reinforce what were public administration responsibilities and how to turn it more economical. At the same time, operational strategies and processes should be "organized around fluid models based on teams, communities and collaborative alliances structured in networks of knowledge," and horizontal processes were needed to better fit citizens' needs – "horizontal networks of inter institutional work or value chains" [12]. According to what was mentioned above about the following steps for public administration, it was predicted that collaborative and interdepartmental technologies will take center stage at public administration future initiatives [12]. It was not possible talking about public administration improvements without mentioning the emergence and innovation on e-government services, a theme that was understood as "vital process for administrative modernization" and as a way to obtain openness and agility for public administration to meet society's needs [12]. Although transversal projects were needed to approach administration can be achieved, only this way will the efficiency and competitiveness necessary for the success of these initiatives be achieved. The use of technologies aims to "increase citizenship, transparency and citizen participation." This is because technologies could be a very important starting point for administrative modernization, but their success depends entirely on the use that citizens and businesses make of it. However, technology is still seen as a critical success point in promoting digital inclusion associated with administrative modernization [12]. It was also very important to take note that an e-government system mostly requires the use of opinion pools to get citizens and companies closer to government [11] feeling that their opinion counts. Besides that, an improvement will only be possible after taking advantage of "the relationship between municipalities and the government" because, nowadays, it "is very scarce" which distances local and central administration [7] which may have caused a lack of communication which directly affected citizens. Customer relationship management systems, which may help, were considered as an important starting point for an improvement in the relationship between the public administration and its citizens. This has been done taking into account that it was possible to draw up a profile of each one managed by tailoring the services to their specific needs. It is a system "where the citizen comes into the spotlight" [1]. At a Portuguese level, through the time, Portugal has presented many administrative modernization strategies as an "introduction of administrative simplification systems," "improving the quality of services provided to citizens by the administration," and " dialogue with the citizen" [7]. Those demonstrated citizens' power inside public administration and how improving the relationship between them and public administration could be crucial to reach a modern public administration. Also, the idea behind using e-government at Portuguese public administration was supported by an assumption that a citizen must communicate with it by different channels and public administration must be responsible to manage an internal workflow to get the answer and return to the citizens as soon as possible [1]. A technological innovation in Portuguese public administration was also a new application called Juntar a Junta, launched in April 2017, and had as the main objective bringing closer parishes and their citizens by an active and aware community [13]. Luís Newton, president of the parish council of Estrela, states that only in 2009 a transversal dynamic implementation with the communities and the citizen was possible, where his involvement began to be valued. The involvement of public administration employees during the changing process facilitated its implementation as well as increased the success rate of it, which was only possible by collecting employee's opinions, which sometimes had included a possible solution to known problems but that were never given an opportunity to share it [14]. However, elderly people felt more and more that they were being left behind and that they did not make part of this new modern world they lived in which represent a big challenge when attempting to modernize public administration as more and more parishes have majorly old populations [15]. It was concluded that any innovation inside Portuguese parishes should always consider that they must end up with a cost reduction for citizens, an improvement in the way they live, or an improvement of the way public entities deliver their services [16] as those are defined as what people were expecting to happen after an implementation of a smart initiative. Regarding local administration, their front office parish competencies are the ones that may help improve the relationship between citizens and parish council and let them be closer to each other. Parish competencies classified as Back office were the ones mentioned in line a), f), g), k), m), o), q), s), t), u), v), x), y), z), aa), bb), cc), ee), ff), hh), ii), kk), rr), tt) and licensing of activities like lotteries sales, car upholstery and noisy activities of a temporary nature that respect popular festivals, pilgrimages, fairs, camp and dances, referred in 16th article of the Portuguese law, 75/2013 from September 12th [17].

2.2 *Smart Cities*

To introduce smart cities, it was interesting to mention that it is known that around 75% of the world's population lives in urban areas and that this percentage tends to grow at least to 80% in 2020. The main objective of governments identified in 2015 was to support this continuous growth by making cities more sustainable, always ensuring the quality of life supported by the implementation of smart cities initiatives [18]. It is also important to reinforce that there was funding for smart city initiatives inside Portugal; it was only necessary that their implementations went according to the specific needs of the citizens as well as have the necessary involvement of parish employees, which was not always the case and therefore leads to the failure of its implementations [14]. In fact, trying to modernize cities around Portugal had as a challenge citizens' involvement in e-services and from the government side an inability to effectively answer to citizens' participation [5]. Although it was recognized that "nothing better than technologies to minimize decision-making arbitrariness," technologies such as "multiple and universal interaction channels, business intelligence, automated reporting and alerting, work flows, mobile technologies" are the future [12]. Smart cities could be a combination of three dimensions that should be perfectly aligned to have success in smart cities initiatives. They are people, community, and technology [18]. Smart City 3.0 is the one related with citizens' co-creation where citizens were involved in initiatives [19]. Regarding technology, it is known that citizens identified advantages of having a website available like its ease of use, its low cost and its lowering the level of bureaucracy inside cities. The major reasons identified to unused e-governmental services by citizens were the absence of necessity to use them and the preference for a personal contact service [16].

By 2014 Portugal had a collaborative platform called smart cities Portugal, where "companies, clusters, universities, R&D centers, municipalities and other economic and social players" acted together to improve smart cities experience inside Portugal [20].

Also about technology and how it is used inside Portuguese public administration, an interesting study made by NOVA IMS students has proved that municipalities provided data to an open-source repository because they were bound to do it in a regular attendance and not because they wanted to share data or even understood the benefits of sharing it regarding the increase of transparency, participation, quality of services, efficiency, and economic development [21].

2.3 *Innovation and Creativity*

It was considered important to study innovation and creativity topics a bit deeper as the truth was that "innovators… need to innovate innovation" [22] every day.

As a consequence, changes around the globe affected the way people do and think innovation. Crowdsourced creativity and software communities are two concepts that emerged during the last decades to accelerate innovation [22].

Although it is known that for an innovation to succeed, innovators needed to be part of online communities to spread the news with all interested parts, only by doing so an idea may succeed because it would be known and understood by a group of people [23]. McKinsey in 2018 revealed that people needed to feel comfortable on sharing their thoughts and also have the possibility to network with others to mature ideas [24]. Searching by mechanisms that helped on stimulating innovation concepts like broadcasting, brainstorming, licensing out, networking, and expert teams were found [25]. Regarding the most common ways to fund innovations, we should mention personal, family, and friends' funding, government grants, debt or equity funding, business angels, venture capital, crowd funding [26], academic partners, customers' or employees' funds if in a corporate context, bankers [27], and so many other ways of get financial help. Venture capitals and business angels were defined as the ones that by having a prototype of what is being innovated are more likely to invest [28]. Co-working spaces also appeared to associate with innovation as they allowed "cost-effective office spaces" with access 24/7 and a community of people who had the know-how about a specific theme/area [29], which may provide access to an "environment of like-minded entrepreneurs, …specialized experts, …high-value networking, and seminars" [29]. Combining design thinking, lean startup, and agile methodologies made it possible to transform an idea into a business solution that better fits the market by delivering the final solution in a step by step manner. What decreased risk level associated with failure after launching new ideas to the market. Also, one of the most important facts of using this combined approach was considered to get people involved in the process, letting them be more fulfilled with their work and proud of the final solution launched [30]. When thinking about carrying forward an innovation accelerator, what people need to consider first is who will be part of the team, because they were mentioned as the essence of an accelerator and that which will make the difference at go, no-go moment [23]. There is a need of a team that involves people from different generations, to get different inputs and perceptions, pursuing a way to meet consumer needs [22]. There are five main elements that characterize an innovation accelerator: first of all there should be a definition of ways of funding created at the end of these initiatives; secondly find "company founders" and recruit people with technical background to be part of the team; after that, define the time window for each group; after having the people and a time window for them to meet, it is time to organize an educational program to advise each team on business/products; and, last but not the least, there should be a "networking program" to join all the teams and advisors to exchange opinions about what was done which may include the final demo day where all innovations that were created are presented to collect insights from people who better know the business [31]. Innovation accelerators were known by including "a fixed-term, cohort-based program, including membership and educational components, that culminates in a public pitch event, or demo day" as well as mentoring, resources, and industry connections and, most importantly, industry connections during these programs [32]. Faster innovation programs like a hackathon differed from innovation accelerators mainly in the fact that they usually did not have the time to let people be familiar with the innovation methodology that needed to be followed, neither changing their mindset according to it. At an innovation accelerator all of these took some time but at the end the results were considered much more disruptive and that better fit core business objectives [33]. The types of sponsors generally involved in an accelerator were universities and corporations [32].

3 Methodology

After some thought about what should be the process to design an innovation accelerator for parishes, to follow a design science research methodology sounds like the best solution to get things done. The choice of this methodology has had into consideration some of its characteristics like the fascination of combining synthesis brought by the design fundamentals and an analytic point of view that came from a scientific background [34]. In this section it will be explained how all the phases of design science research methodology will be used during this study and what will also be the tasks inside each of them. It is possible to count with a help of the following model to explain it: As shown above, this study started at design objectives and solution stages because it was driven by an objective-centered solution which is to bring citizens/companies closer to parishes using an innovation accelerator. So, to start, there was the need to define objectives and a solution around the main problem identified in the previous paragraph, to get to know which areas the basis of this study should have according to requirements that fitted the solution's goal. Design and development stages involved research around three main themes: public administration, its current technological state, and what were the main responsibilities of a parish council in Portugal, smart cities, main concepts and initiatives that were already being done, and innovation/creativity and existing methods to boost it. This stage was split into two tasks, the first one explained above and the second one that involved the creation of an innovation accelerator according to what was researched and set as requirement by defining what is essential to get smartness inside parishes. The last stage, inside the scope of this study, but one of the most important one, was the evaluation of the output artifact that was done with the help of a focus group meeting to get feedback from participants about what their impressions were about the proposed artifact and how they saw the possibility to implement it based on the context they have on parish council's reality. The moderator was Dr. Emanuel Costa member of executive committee from the Lisbon Metropolitan Area. Participants were three Drª. Fernanda Marques, director of economical and local development department; Dr. Bruno Martinho, municipal director of economy, innovation, and communication; and last but not the least José Ricardo Dias Martins, president of Costa de Caparica parish council. Those participants were chosen by the moderator as he better knew people from Lisbon Metropolitan Area Parishes and Municipalities, this chosen were made according to our pre-requirements that were to have at least one people from innovation, one people from a parish and one people from municipality in order to bring knowledge about other parishes as well. Costa de Caparica was considered an interesting parish as it already follows a longest path regarding innovation, and we wanted to evaluate how aligned the path they followed was with the proposed framework. Also, we wanted to know how the framework fits their next steps on innovation. A person from a municipality was considered mandatory as it has the context not only from urban parishes but also from rural ones.

4 A Model of an Innovation Accelerator for Smart Parishes

After studying Portuguese public administration, smart cities, and innovation, it was possible to have a clear notion on what must be included in the definition of the main concept behind this study: a smart parish. In fact, it was understood that a smart parish is the one that wants to improve the relationship between citizens and parishes by promoting the dialogue between them, including more citizens in the initiatives that happen inside their parish, listening to their opinions, and making them feel like their opinion counts. Also, a smart parish is the one who gives training to parishes council's employees, for them to be more focused on providing a citizen-centered service; the focus are the citizens and not only the process and the bureaucracy involved, making citizens closer to parishes. Giving IT training to citizens to take advantage of information technologies and promoting digital inclusion is also part of smart parishes' mission. In general, a smart parish must improve the way people live in it by involving who lives and/or works there. At the scope of this study, an innovation accelerator for parishes was proposed, for them to become smart parishes; according to the assumptions described above, it was proposed the following conceptual framework. The main goal of the conceptual framework proposed is giving citizens a better quality of life by improving all communication channels between them and parish councils, to get them closer, and also by letting them have a voice inside the parish by sharing their concerns and thoughts and giving ideas to improve their daily lives. Who better than the people who work/live inside a parish have an idea on what is good or bad in it? In Fig. 2, it is possible to see all the main concepts of the framework and who are the participants of each one.

Based on this, it was considered that for a parish to become smarter, changes needed to start from inside the parish council: by giving attention to their employees, listening to their ideas, re-organizing processes to be able to respond to citizens in a more effective and efficient way, and also involving them in the modernization of services given to citizens by training them on innovation to increase the chances for innovation initiatives to succeed and be accepted inside the parish (Fig. 3). Also, citizens needed to feel they belong to a parish which listens to their thoughts and that improves their way of living and this is only possible by letting citizens network and brainstorm with each other to create new and disruptive ideas that have real pains behind with a help of mentors from specific areas and from the parish council itself. Industry connections are also needed components in the proposed accelerator as they will catch connections from companies and parish council to find some fund, mentorship, and/or physical space to help in implementing and guiding innovations. An improvement of parish council's website was considered a mandatory component as, after studying parish councils' competencies, it was found that the front-office ones, that directly influence the relationship between citizens and parishes, had a lot to do with providing information to citizens and forms for them to fill in. Also, a website improvement is needed in order to be possible to have a page in which people could submit their ideas and later require funding for them. Because of this, a funding component was also needed at this parish accelerator where citizens (business angels), companies, or even the parish council with the help from government would help take some innovations to a next level. Considering this wave of innovations, concerns came up, people from parish council needed training to know how to deal with innovation, and citizens needed training on how to make use of the e-services they have at their disposal, how to implement their ideas, and how to convince people to invest on their ideas taking them to a next level and training to promote digital inclusion. Based on that, education was also considered a mandatory component inside this framework. Co-working spaces had already been implemented inside the context of smart cities and had been a success for people who want a place to work or study without being at

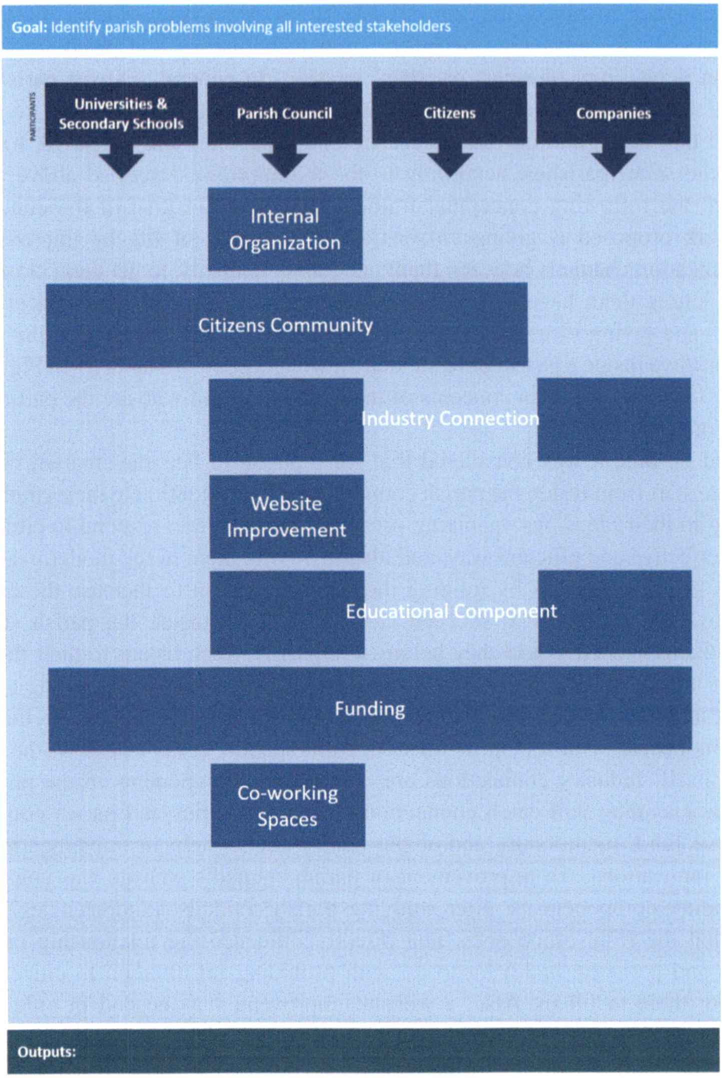

Fig. 2 Framework components

home or paying a very high income. Also, as one of the outputs of this framework is to generate ideas, people need a place to do it and to meet with other citizens or even with some investor. So, co-working spaces were a need in the context of the proposed framework. It was considered that the implementation of the proposed framework should start from inside the parish councils to the citizens. So, in general terms, first the parish council will need to rethink and re-organize the way they do things and the services they provide and only then improve communication

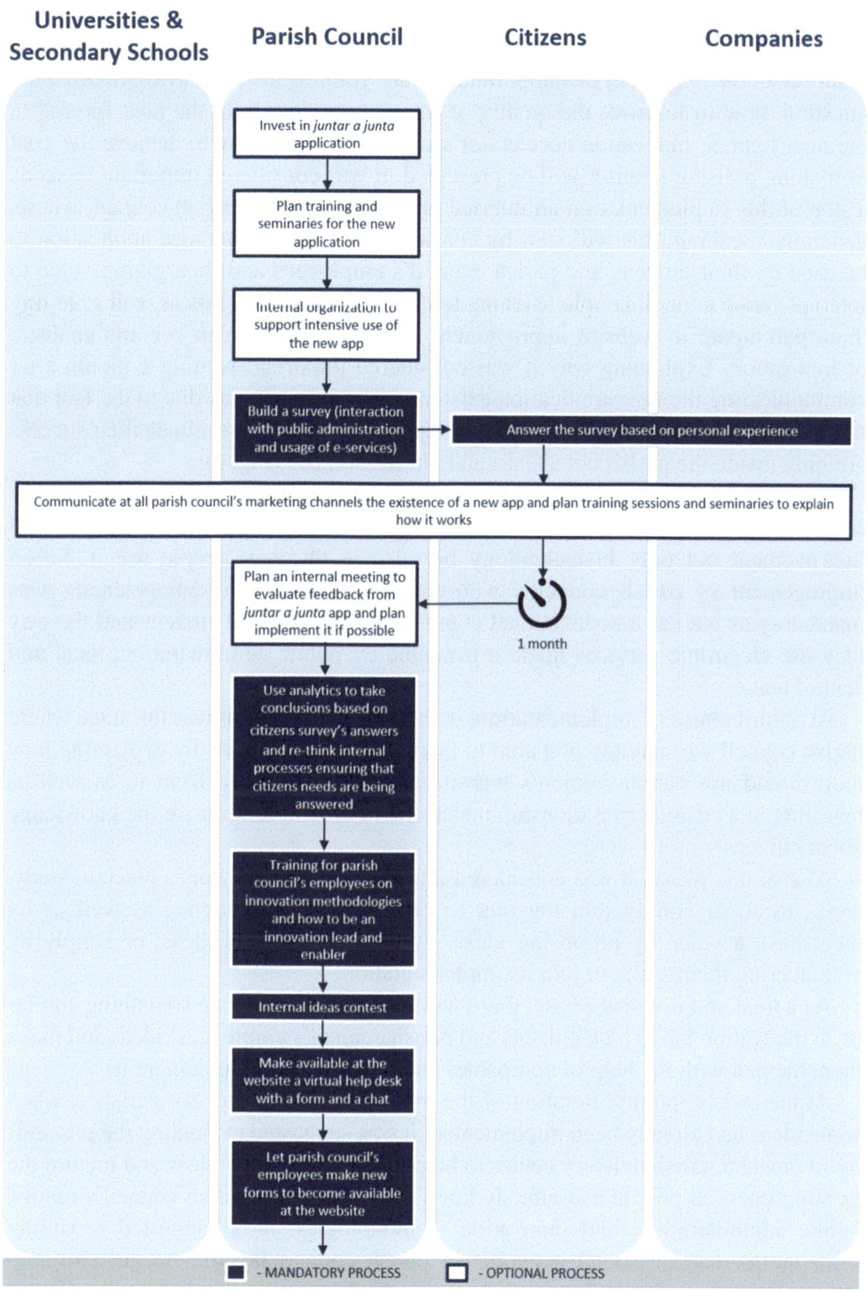

Fig. 3 Framework Implementation Phase 1

channels between citizens and parish council giving voice to citizens and, if possible in the context of the parish, promote an innovation lab where ideas will become reality with the help from mentors, funding, and training always having in mind the question: how to improve the quality of life of your parish, in the near future? In the next figures, innovation accelerator's implementation flow to achieve the goal of making parishes smarter will be presented. It was considered important to set as a step of this implementation an internal organization of the parish council, where, generally speaking, this will start by considering investing in a new application to be used by their citizens and parish council's employees and then giving voice to internal parish council people listening to their ideas and thoughts as well as letting them participate in website improvement to encourage them to become enablers of innovation. Explaining why it was considered important waiting 1 month after communicating the new application existence to citizens, it was due to the fact that it was considered that only after this time will it be possible to evaluate their success not only inside the parish but also as an help for citizens (Fig. 4).

After improving things internally, it was considered that the second phase of implementation of the proposed framework should include an investment and improvement not only in technology but also in the way people use it. So an improvement of parish council's website aligned with their competencies was mandatory as well as an educational component that included citizens and the way they use electronic services made it available by public administration, local and central one.

At a third phase of implementation, it was considered that it was the stage where parish council was already prepared to involve citizens. It starts by explaining how they should use parish council's website and take advantage from it, as well as investing in a customer relationship management system to improve the knowledge about citizens.

Also at this phase, it was considered important to give citizens a place to study, work, meet, or simply join together to exchange some thoughts, as well as let them have a voice by proposing ideas, by voting in other's ideas, or simply by volunteering themselves to join its implementation.

As a final and optional phase, there was a need to incorporate something similar to an innovation lab to pick citizens and parish council's employees' ideas and make them happen with the help of companies and universities to implement it.

At the end of the first iteration of the innovation accelerator for parishes, when some ideas had already been implemented, it was suggested evaluating the possibility to create a parish delivery center to help implement future ideas and mature the existing ones, as people had already knowledge about the parish council's reality, public administration, and innovation methodologies. It is suggested recruiting someone to manage social networks of parish council in order to get closer to citizens, getting their attention on what is being done, as well as influence external opinions with the help of parish influencers (Fig. 5).

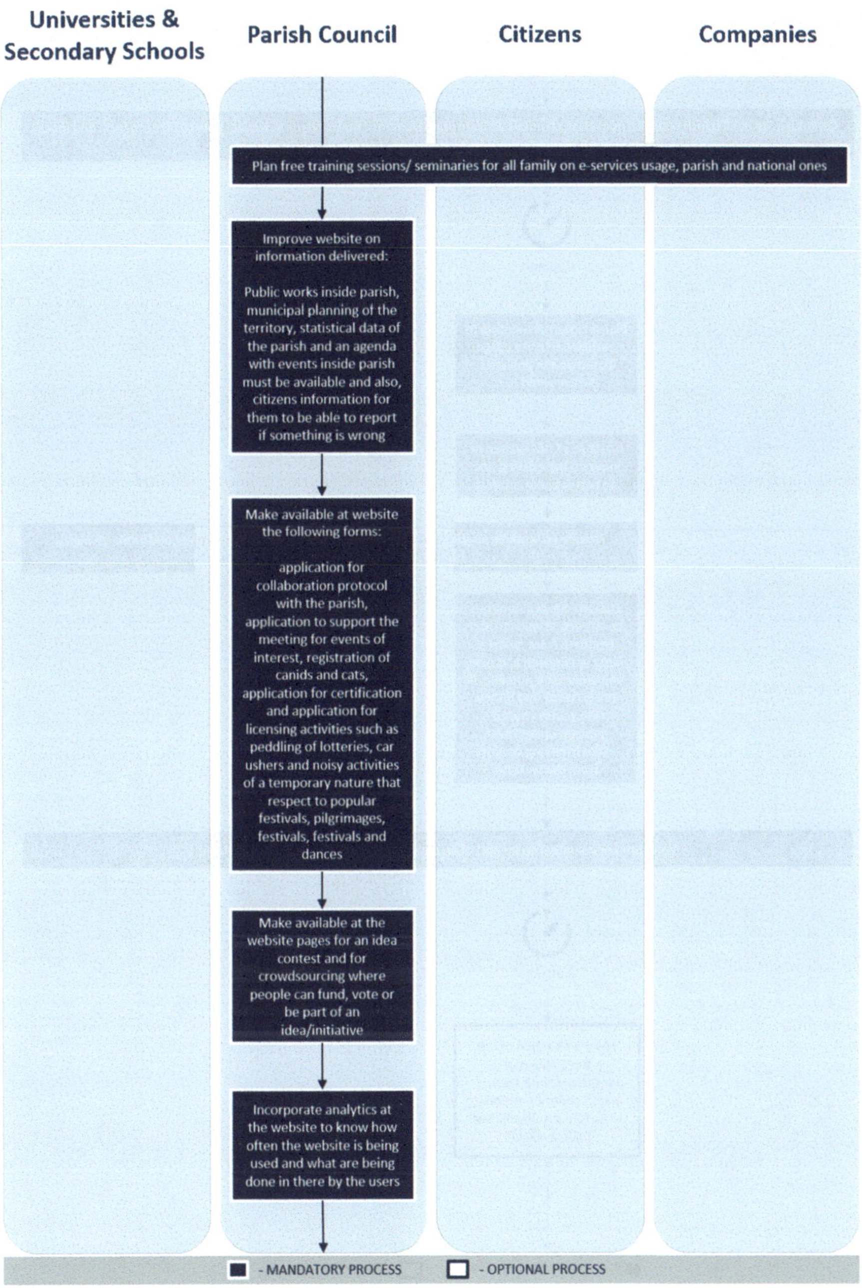

Fig. 4 Framework Implementation Phase 2

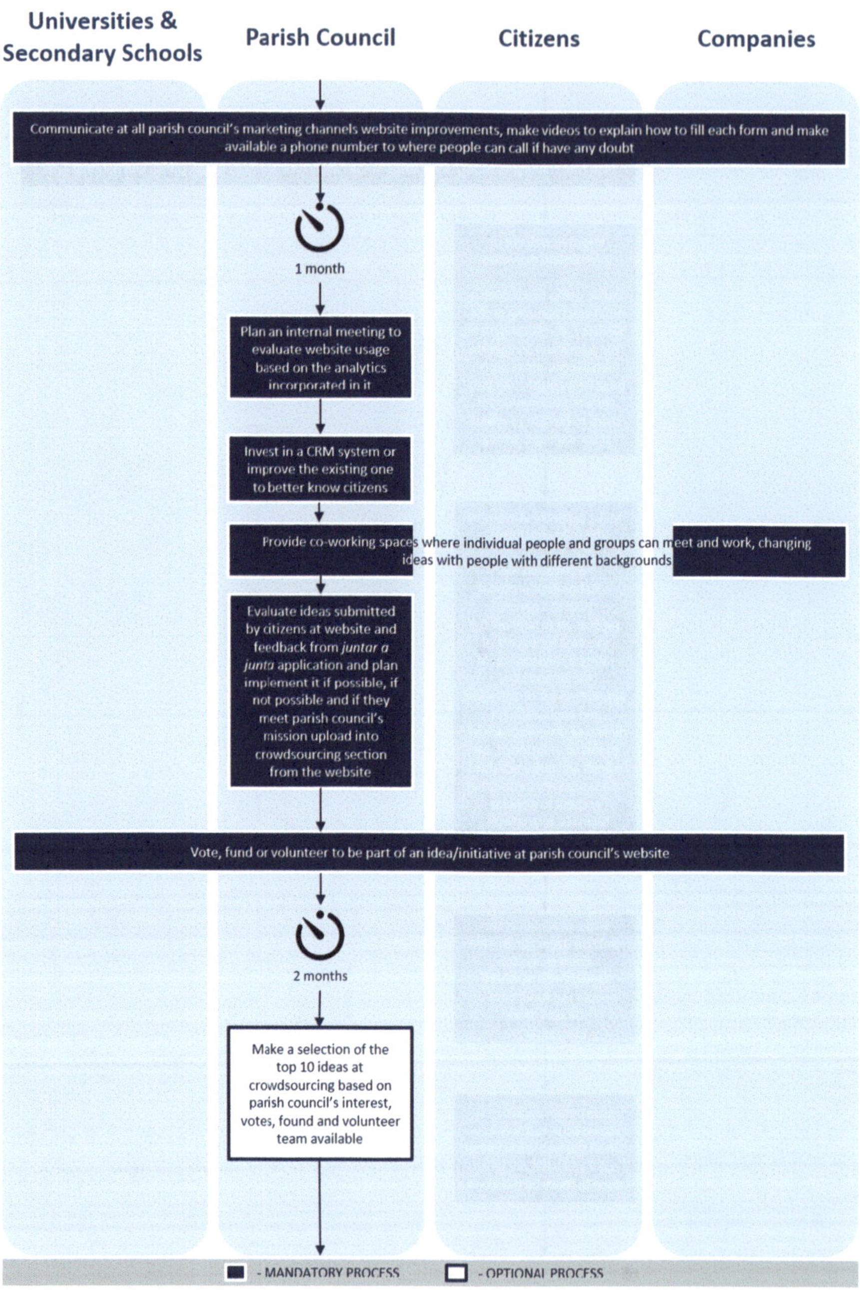

Fig. 5 Framework Implementation Phase 3

5 Discussion

In this section, three analysis will be made, regarding utility, viability of implementation, and improvements based on the answers from validation phase. Then a general evaluation to the proposed framework will be done based on the previous analysis. So, with respect to the proposed conceptual framework and with regard to the utility of it, the three participants of the focus group meeting agreed that the proposed accelerator is very useful. It was considered, by the participants, as indispensable and seen as a way to agglutinate citizens and public administration letting citizens participate more and be more involved, although they express their concerns regarding the need of adapting the process to each parish's reality enforcing the idea that one size does not fits all and it depends on the parish itself and on its stratum. Also, and still regarding the utility of the proposed accelerator, it was a shared idea that this could be amazing, specially, as a way to force public service itself to be upgraded and its skills in some subjects. With the adoption of this framework, it was considered that it is important to get parishes closer to what is considered the state of the art. Internal parish improvement is a progress that is already very important, independent of the improvement on citizens' participation, as it was considered easier to get citizens involved.The utility of the proposed accelerator was also evaluated in a way that the types of communication channels included in it are fundamental nowadays, so it was a positive point regarding the utility of the accelerator. When talking about observations on what was proposed and an evaluation on how viable it is, things were discussed like the need to get help from city council on communication phases as a premise for frameworks' implementation as it was proposed to be one of the starting points of the framework near citizens, the need to have a solid network between city council and parish council, and the need to clarify with citizens what were the competences of city councils and the ones from parish councils as today they do not know yet to whom they should ask things on specific subjects. As a way to get a more sophisticated framework, it was proposed that it is really fundamental to look into public service levels, analyze also the sophistication of the population itself, and last but not the least evaluate the levels of participation from citizens. Those three gaps need to be evaluated and improved even before starting frameworks' implementation as those will influence the success of it. Although and besides all the previous observations, the framework was considered not as the perfect model for parishes but as a very advanced one following the needs of parish councils today, as the one who perfectly fits what is being done at city council's level. Also, as a positive point to the possibility of implementing this framework, it has the ability of parish councils to adapt, as they are constantly changing and adapting to new competences that city council delegates on them. Another observation made about the proposed framework was that it was considered mandatory involving people from all ages in it and using different communication channels. By combining this two, it is

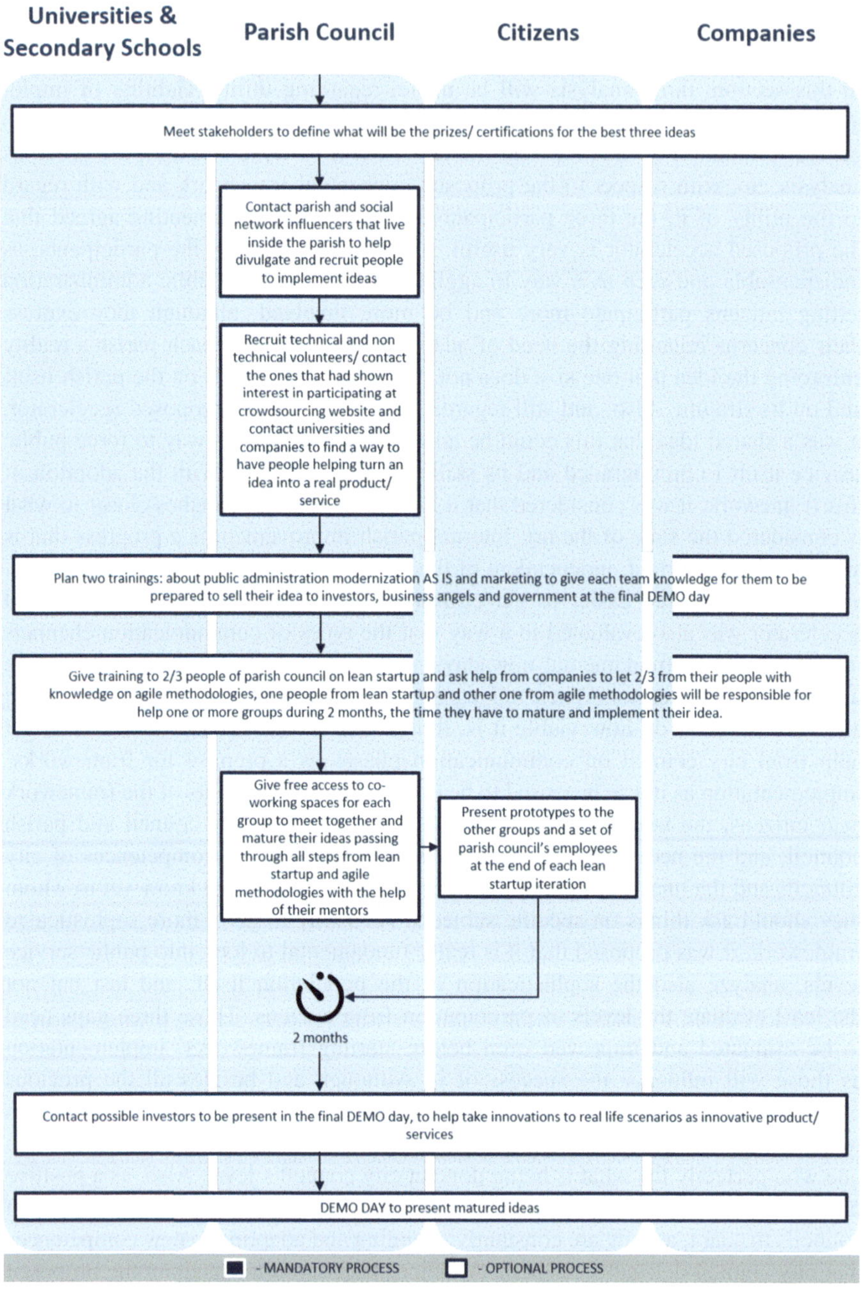

Fig. 6 Framework Implementation Phase 4

possible to get mixed and progressive models of dematerialization. To conclude, it was considered viable the proposed framework as it would contribute to define a procedures matrix of citizens' relationship. As criticism and suggestions for improvement, all participants agree that it would only be possible to evaluate with a practical application of the proposed framework and tests around it. In general, the proposed framework fulfills the needs of parishes in bringing parishes closer to citizens by letting them have an active voice and also in improving public administration services and skills. So, the process included in this framework is fundamental not only for today's reality of parish councils but also to guide them to the future. It is understood that it has utility, although it needs some adaptations to different parishes realities, a more detailed communication phase where city councils must be an intervenient, and the possibility to have a way to evaluate the three gaps of sophistication mentioned above as a way to adapt the framework to each parish reality (Fig. 6).

6 Conclusions and Future Work

During this work subjects like public administration (national and international), smart cities, and innovation and creativity were studied as a starting point to be able to design a framework that later on was validated by a focus group meeting by people with knowledge on public administration, especially on local public administration subjects. An improvement to the proposed model was considered as fundamental; after validating the framework, it was concluded that some processes must include as intervenient the city council as it has the resources to help take parishes to the next level. A limitation of this work was the fact that it was not possible to validate this framework by practical application; it was considered that it should be done in the future in order to improve it based on a practical application in a real-life scenario. However focus group was composed by people with knowledge on different parishes in order to enrich the inputs. As the last consideration for future work on this subject of smartness inside parish, it is considered important to evaluate the option to improve the proposed framework extending ideas contest and innovation lab to several parishes instead of implementing it in only one parish at a time. However and besides all the future work identified previously, it is important to mention that predefined objectives of this work were achieved. By the application of the proposed framework and according to what was the feedback collected at validation phase, parishes will become smarter by the implementation of the proposed innovation accelerator.

References

1. Cardoso, J. (2014). Sistemas de informação para a modernização administrativa. 11.° Encontro de Arquivos Municipais. Esposende: BAD – Bibliotecários, arquivistas e documentistas.
2. Royo, S., & Yetano, A. (2015). "Crowdsourcing" as a tool for e-participation: Two experiences regarding CO2 emissions at municipal level. *Electronic Commerce Research, 15*, 323–348. ISSN: 1389-5753.
3. Walters, D. (2011). Smart cities, smart places, smart democracy: Form-based codes, electronic governance and the role of place in making smart cities. *Intelligent Buildings International, 3*, 198–218. https://doi.org/10.1080/17508975.2011.586670.
4. Panagiotopoulos, P., Al-Debei, M., Fitzgerald, G., & Elliman, T. (2012). A business model perspective for ICTs in public engagement. *Government Information Quarterly, 29*, 192–202. https://doi.org/10.1016/j.giq.2011.09.011.
5. EUROCITIES. (2017). Smart city baseline report Lisbon. EUROCITIES.
6. Governo quer freguesias com mais meios e mais competências. (2018). https://www.portugal.gov.pt/pt/gc21/comunicacao/noticia?i=governo-quer-freguesias-com-mais-meios-e-mais-competencias. Last accessed 26 Jan 2018.
7. Rodrigues, C. (2011). Governação de organizações públicas em Portugal: A emergência de modelos diferenciados. Edições Pedago, LDA, Mangualde.
8. Grimmelikhuijsen, S., & Meijer, A. (2017). Open data for democracy: Developing a theoretical framework for open data use. *Government Information Quarterly, 34*, 45–52. https://doi.org/10.1016/j.giq.2017.01.001.
9. Silva, C., Lamas, M., Castro, R., Silva, S., & Rocha, Á. (2018). Governo Electrónico nas Juntas de Freguesia: Situação na Região do Minho.
10. Almeida Henriques: "Smart cities" precisam de apoios. (2017). http://www.jornaldenegocios.pt/negocios-iniciativas/cidades-inteligentes/detalhe/almeida-henriques-smart-cities-precisam-de-apoios. Last accessed 30 Mar 2017.
11. Helbing, D., & Balietti, S. (2011). How to create an innovation accelerator. *The European Physical Journal Special Topics, 195*, 101–136. https://doi.org/10.2139/ssrn.1753796.
12. Mateus, J. C. (2008). O governo eletrónico, a sua aposta em Portugal e a importância das tecnologias de comunicação para a sua estratégia. *Revista de Estudos Politécnicos, VI*(9), 23–48. ISSN: 1645-9911.
13. Nova app faz ligação direta às juntas de freguesias. (2017). http://visao.sapo.pt/actualidade/sociedade/2017-10-30-Nova-app-faz-ligacao-direta-as-juntas-de-freguesias. Last accessed 30 Oct 2017.
14. Primeiro desafio para "smart cities" é de back-office. (2017). https://www.computerworld.com.pt/2017/05/05/primeiro-desafio-para-smart-cities-e-de-back-office/. Last accessed in 2018.
15. Azevedo, C. (2013). Tecnologias e pessoas mais velhas: Importância do iso e apropriação das novas tecnologias de informação e comunicação para as relações sociais de pessoas mais velhas em Portugal.
16. Ferreira, N. C., Ferreira, F. A., Marques, C. S., Ilander, G. O.-B., & Çipi, A. (2015). Challenges in the implementation of public electronic services: Lessons from a regional-based study. *Journal of Business Economics and Management, 16*(5), 962–979, ISSN: 1611-1699.
17. Regime Jurídico das autarquias locais. (2013). http://www.pgdlisboa.pt/leis/lei_mostra_articulado.php?nid̄1990&tabela=leis. Last accessed in (2018).
18. Albino, V., Berardi, U., & Dangelico, R. M. (2015). Smart cities: Definitions, dimensions, performance, and initiatives. *Journal of Urban Technology, 22*(1), 3–21. https://doi.org/10.1080/10630732.2014.942092.
19. The 3 Generations Of Smart Cities. (2015). https://www.fastcompany.com/3047795/the-3-generations-of-smart-cities. Last accessed in 2018.
20. INTELI. (2014). Smart cities Portugal roadmap In brief. INTELI.

21. Neto, M. D., Rego, J. S., Neves, F. T., & Cartaxo, T. M. (2017). Smart & open cities: Portuguese municipalities open data policies evaluation. In *12th Iberian Conference on Information Systems and Technologies (CISTI)*. https://doi.org/10.23919/CISTI.2017.7975912.
22. It's a fast, VUCA world: the trends and forces shaping innovation. (2014). https://www.theinovogroup.com/its-a-fast-vuca-world/. Last accessed in 2018.
23. Six Lessons for Corporations Building Innovation Accelerators. (2013). https://www.innosight.com/insight/six-lessons-for-corporations-building-innovation-accelerators/. Last accessed 29 Aug 2018.
24. Barsh, J., Capozzi, M. M., & Davidson, J. (2008). Leadership and innovation. *McKinsey Quarterly, 1*, 37–47.
25. 5 Mechanisms to Promote Innovation. (2018). http://www.esadeknowledge.com/view/5-mechanisms-to-promote-innovation-174880. Last accessed 5 Jan 2018.
26. Sources of Innovation Funding. (2018). https://www.business.qld.gov.au/running-business/growing-business/becoming-innovative/funding/sources. Last accessed 8 Aug 2016.
27. Top 10 Sources of Innovation: Why Your Accelerator Barely Makes The List. (2018). https://innov8rs.co/news/top-10-sources-of-innovation-why-your-accelerator-barely-makes-the-list/. Last accessed 2 Sept 2018.
28. Business Angels. (2018). https://www.innovationpolicyplatform.org/content/business-angels. Last accessed 7 Dec 2018.
29. Coworking Space vs. Startup Incubator vs. Accelerator: Which is Best? (2017). https://www.rocketspace.com/tech-startups/coworking-space-vs.-startup-incubator-vs.-accelerator-which-is-best. Last accessed in 2018.
30. The Difference Between Design Thinking, Lean Startup, and Agile. (2017). https://medium.com/@SteveGlaveski/the-difference-between-design-thinking-lean-startup-and-agile-5cf07b117562. Last accessed in 2018.
31. Fernandes, S. (2016). Innovation accelerators as entrepreneurial and interdisciplinary engines: The Portuguese case. *Journal of Spatial and Organizational Dynamics*, 213–228. ISSN: 1647-3183.
32. Jennifer Auer, C., & D'Ippolito, M. (2014). *Innovation accelerators: Defining characteristics among startup assistance organizations*. College Park: Small Business Administration – Office of Advocacy. https://doi.org/10.13140/RG.2.2.36244.09602.
33. How to design your own innovation accelerator: a checklist. (2015). Obtido de Board of innovation. https://www.boardofinnovation.com/blog/2015/10/01/a-checklist-for-designing-your-own-innovation-accelerator/. Last accessed in 2017.
34. Baskerville, R. L., Kaul, M., & Storey, V. C. (2015). Genres of inquiry in design-science research: Justification and evaluation of knowledge production. *McKinsey Quarterly, 39*(3), A1–A9.

A Novel Intelligent WEB Application on Refrigerator Systems

Hüseyin Gürüler and Gürcan Çetin

1 Introduction

Network-enabled devices (NED) offer exciting connectivity to household devices. With the rapid expansion of the broadband Internet, remote controlling and monitoring of NED became easier. On the other hand, many devices that enable user-machine interaction, such as the Internet of Things (IoT), big data, cloud computing, robotics, sensors, etc., continue to be adapted with the concept of Industry 4.0. These concepts are closely related to receiving data from heterogeneous devices, processing huge amounts of distributed data, and interoperability [1].

IoT-enabled smart home devices have been designed and developed by many academics and vendors. Wang et al. have published [2] a study on IoT-based system for intelligent homes. In another study, Su et al. [3] also introduced an IoT-based smart home, where communication and collaboration between individuals and devices are achieved via short message service (SMS). An alternative solution in the implementation of smart homes has been offered by Cheuque et al. [4] using Raspberry Pi. These products can communicate with environmental units through various communication protocols such as Ethernet, Bluetooth, wireless sensor network (WSN), and radio-frequency identification (RFID) [5]. These smart devices, which are capable of various communication and computation, are available in the market. At this point, IoT serves as an umbrella that encompasses a heterogeneous network of objects [6]. Moreover, assistive IoT technologies in smart homes are strong tools that resolve the social isolation of people with disabilities by designing smart home automation platforms and enhancing existing device features to improve living conditions [7]. Welbourne et al. introduced [8] an RFID-based

H. Gürüler (✉) · G. Çetin
Information Systems Engineering, Mugla Sitki Kocman University, Mentese, Mugla, Turkey
e-mail: hguruler@mu.edu.tr; gcetin@mu.edu.tr

M. M. Banat, S. Paiva (eds.), *Smart Technologies for Smart Cities*,
EAI/Springer Innovations in Communication and Computing,
https://doi.org/10.1007/978-3-030-39986-3_2

application for blinds that consists of a search engine and a web platform to find objects. Firstly, personal objects are labeled with RFID tags according to the object-related metadata, and then this information is managed through a web app at home.

Another triggering factor to integrate the smart home and smart device concepts into daily life is the widespread use of mobile applications and web-based platforms due to their lower prices. One of such systems is reported [9] that has the home top view, live camera streaming, and schedule and rules manager. Another mobile application [10] controls and monitors parameters of an enhanced smart oven through Wi-Fi. Another reported system [11] for disabled and elder persons to control many features of a smart home such as turn on a light or switch off a fan.

On the other hand, a web-based platform supports user-to-machine or machine-to-machine interaction over the Internet. WEB technology has been transferred into devices. Guruler et al. [12] designed an XML-based microcontroller system to monitor the actual currency of the foreign exchange data of the central bank. Yüksekkaya et al. [13] conducted a home automation system using speech recognition through GSM communication and the Internet.

The combination of IoT, mobile applications, and web-based platforms along with advanced communication and sensing systems has created a variety of new potential applications in the area of smart homes. This study answers the question of how all these new potential developments in smart homes can be integrated into an existing product cheaply. This chapter proposes a new smart mini fridge that can be used in homes, hotel rooms, or offices and can be controlled/monitored by WEB-GSM technologies. It monitors the number of products left in the fridge; thus it can provide instant control, stock follow-up, automatic invoicing, more effective use of human resources in hotel management, and customer satisfaction.

Another target group of this project is the people who have summer houses or farm/village/highland houses. These homes are usually second homes and don't have long-term dweller. Because of this, the homeowners, especially in highland houses, should consider the products they put in refrigerators that need to be cooled against any problems such as a power failure. The smart refrigerator developed in this project can send information to owners via SMS in case of a power failure. It is aimed to drive the smart home systems up to the highland houses on the way to disseminate.

The main targets are to keep the cost at least as possible during the commercialization of the project. In this way, the study also aims to show that this is a need for people who think that smart home systems are luxury and that these technologies can be obtained in a cheaper way not only for high-income people but also for all income groups.

The project is also intended for the widespread use and an OS-independent manner.

The main contributions of the study are listed as follows:

- The study proposes a new and cost-effective mini-refrigerater allowing user-machine interaction through a remote monitoring infrastructure.

- By using embedded WEB pages and GSM SMS modules in the developed product, the information about device and product stocks are send, owners are informed when necessary, and they can perform a distant query.
- While this work focuses on the development of a smart refrigerator, it also contributes to our understanding of the location of smart devices within smart homes and smart cities of the future.

The chapter is structured as follows: Literature review is given in Sect. 2 to provide a summary and assessment of the studies on the development of smart refrigerators. In Sect. 3, there are detailed materials and methods employed in this study followed by the working principle and algorithms of the system. The conclusions of the study are given in Sect. 4.

2 Literature Review

Smart homes are cyber-physical systems that allow users to monitor and control different home appliances remotely and intelligently to meet their comfort or safety expectations [14]. It is recognized in many commercial and academic environments that smart homes and devices are the most accepted and rapidly expanding technology among all IoT applications. Estimates indicate that the global smart home market will grow by 12.02% annually between 2018 and 2024 that will make a market volume 151.4 billion USD. The main effects of this spread are the increase in the number of Internet users, adoption of mobile and smart devices by the society, the awareness of healthy living and comfort in developing countries, consequently allocating more budget to such devices, and understanding the importance of monitoring and remote-controlled systems [15]. These effects also recently led to a significant increase in the smart home portfolio. There are many reports on devices such as fuzzy logic-based smart oven [16], smart microwave oven with image classification [17], secured smart digital door lock [18], smart curtains [19], smart light [20], smart fire and security systems [21], smart TV [22], smart washing machine [23], etc.

The smart mini-refrigeraters, on which this chapter focuses, are another smart home product. The refrigerators are a storage area holding food and drink to keep it fresh and healthy for a long time at a certain temperature. These sophisticated devices have important place in our lives together with adding many more features and enhanced capabilities. While home and office environments are becoming more and more intelligent and user-friendly, it is inevitable to introduce new features to industrial products used in these environments. Efforts to develop refrigerators that are a part of our daily life as more intelligent and user-friendly are becoming more and more widespread in both industries and in academic research.

In 2000, LG industrially produced a refrigerator with an Internet connection for the first time. This product, called Digital DIOS, includes a TFT-LCD screen and TV over LAN, providing information such as data memos, video messaging, schedule management, webcam functions, internal temperature, and freshness of stored products [24]. In 2013, Samsung launched the T9000 prototype with an LCD display. The prototype includes Google Calendar, Evernote, and Samsung's smart home applications which are accessible via the Internet. In 2016, Samsung introduced the new-generation smart refrigerator with CES 2016 that includes a 21.5-inch touch screen and a built-in camera. Thanks to this product, users have been able to manually control the snapshot image of the inside of the refrigerator with a smartphone [25]. Although large refrigerator manufacturers will continue to produce smart refrigerators, the desired sales have not been achieved due to the high prices, the fact that many functions are seen as unnecessary or complicated by the users, and the reluctance of the users to replace old refrigerators immediately.

When the smart refrigerator technology is evaluated in terms of academic studies, it is seen that many studies are carried out in parallel with smart homes according to the aims of the ability to be gained to refrigerators. In the studies on food management in smart refrigerators, Singh and Jain [26] have developed an RFID-based robotic system to effectively manage and accurately position food stored in the refrigerator. The collected data was transferred to a mobile application via ESP8266 Wi-Fi module and presented to the users. In the method presented by Qiao et al. [27], the stock information of the food stored in the refrigerator is presented to the users through the cloud data service platform via a wireless connection. Users are informed online 1 week before the shelf life of the food. Another cloud-based smart refrigerator was developed by Wu and Chuang in [25]. The proposed system includes a Raspberry Pi module, sensors, and two (internal and external) cameras. The internal camera provides information to the user about the food inside the refrigerator, while the external camera captures images of the food brought close to it and is used to enter stock information about them. Edward et al. [28] designed a system including Raspberry Pi 3, NodeMCU, and a barcode scanner for product tracking. In addition, the ordering information of the user was kept in the developed system, and integration with the automatic ordering system was ensured.

There are also studies that send error/warning notifications about the refrigerator in addition to the food stock in smart refrigerators. Mahajan et al. in [29] developed capabilities such as giving warnings when the odor in the refrigerator exceeds the threshold, when hot products are placed in the refrigerator, or when excessive weight is detected on the shelves. In another study [30], the features of monitoring humidity, gas, and temperature values and receiving alerts were added. Zhongmin et al. in [31] have also developed Bluetooth-based mobile application to enhance the human-computer interaction by displaying the food condition, temperature, humidity, air quality, and pressure sensors of the refrigerator with various sensors.

One of the main functions of the smart refrigerator, with minimal effort, is to keep an inventory list of food items that can be purchased as soon as their shelf life is full. Luo et al. in [32] focused on how smart refrigerators can offer good nutrition recommendations for human health. In the study, by managing food stored in the refrigerator, users are informed about food that has a shelf life and are advised about their cooking methods. In another study, Shweta [33] stated that artificial intelligence can be used in combination with image processing techniques to determine the freshness of food in the refrigerator. With the cameras placed in the refrigerator, vegetables and fruits such as cabbage, apple, pumpkin, and broccoli were determined with 96.55% accuracy according to their color, shape, texture, and size, and food was monitored with images taken daily. This study suggests that users can be informed via SMS by using smart refrigerator.

We have designed a low-cost smart mini-refrigerater for smart homes. Because commercially produced products are expensive, users do not want to buy a new smart refrigerator at a higher cost when they have a refrigerator. However, the integration of the developed system with the existing refrigerators can be easily achieved. Moreover, no image processing features were added to the developed system. The main reason for this is to keep the price lower for the end user. Image processing technologies require CPUs or GPUs that require high processing power to recognize products. Furthermore, the system does not work effectively when the camera angle and camera image are blocked. On the other hand, many smart refrigerators are designed to be monitored or controlled only by mobile applications. Therefore, they only used one communication protocol such as Bluetooth, ZigBee, and Wi-Fi. In case the communication with the device is broken, the information flow about the smart refrigerator stops. In our device, that can be monitored via the embedded web pages as well as any mobile network.

3 Methodology

Human-machine interaction was achieved in this project, thanks to a client-server architecture. With built-in mobile and WEB modules, the system enables individuals to access product information in the refrigerator from anywhere.

The system architecture can be divided into three parts: the operation principle, hardware, and software.

The hardware of the mini-refrigerater system (Fig. 1) consists of a microcontroller card with a built-in Ethernet module, GSM modem, I/O card, LCD card, power supply card, sensors, and accumulator. The accumulator provides energy to the electronic system when the refrigerator is powered off. It was preferred to detect the temperature of the medium (NTC). In the project, sim900 as GSM module and PIC18F67J60 microcontroller with built-in Ethernet module (Olimex) were used.

3.1 The Operation of the System

The I/O card detects the presence and number of the products in the refrigerator with microswitches and transfers the collected data to the microcontroller card. Product information is stored in the temporary memory of the microcontroller and interpreted by the microcontroller. This information can be accessed via GSM or the Internet which can work independently with each other. This will allow access in

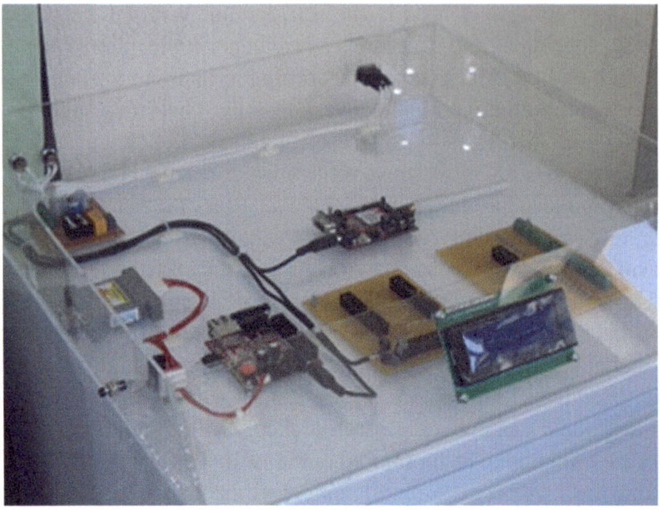

Fig. 1 The system hardware

different situations such as sectoral and personal use. For this purpose, with a mode switch placed in the system considering corporate or individual use, information can be accessed only by mobile phone queries or via the Internet only.

Once the quantities of the products in the mini-refrigerater are obtained by sensors, the necessary data is transferred to the related environments with the open-source microchip TCP/IP stack library, a collection of codes that connects to the subnets or Internet via various Ethernet modules. They can also be considered as a framework library. The Ethernet module is equipped with a PIC18F67J60 microcontroller that makes it possible to connect the microcontroller with the Internet. Within the project, an HTML-based web page was designed, and these HTML codes were loaded into the microcontroller memory. This web page can be viewed using a web browser. The IP address of the device can be typed into the address bar to access the HTML-based web interface integrated into the microcontroller. This web page provides information on products in the refrigerator and freezer as well as the cooler temperatures.

There is SIM900 integrated by SIMCom which is specially designed for GSM/GPRS modems. Communication between microcontroller and GSM modem was provided by using AT (Attention) commands. GSM query can be made by SMS or phone call to the SIM card number in the GSM module attached to the refrigerator. After the sensors data is transferred to the microcontroller board and interpreted by the software inside the microcontroller while the product information is sent to the GSM module. This information is transmitted by text message to one or more mobile phone numbers defined in the system. Thus, users can find out which products are in the refrigerator.

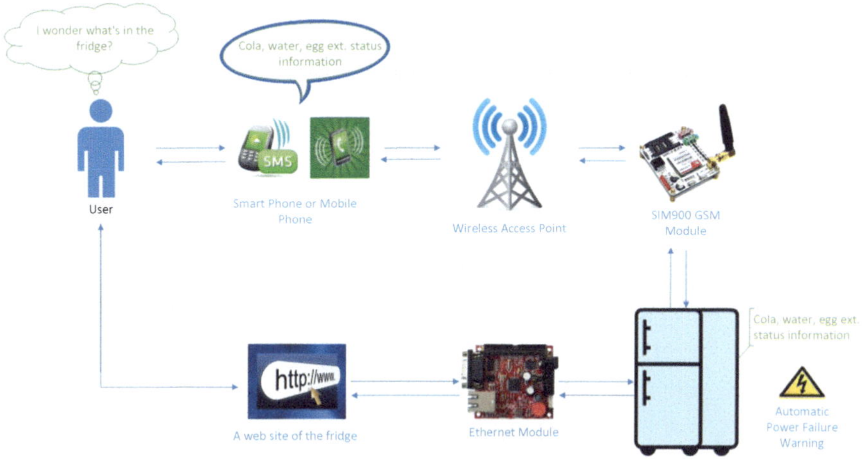

Fig. 2 Operation of the system

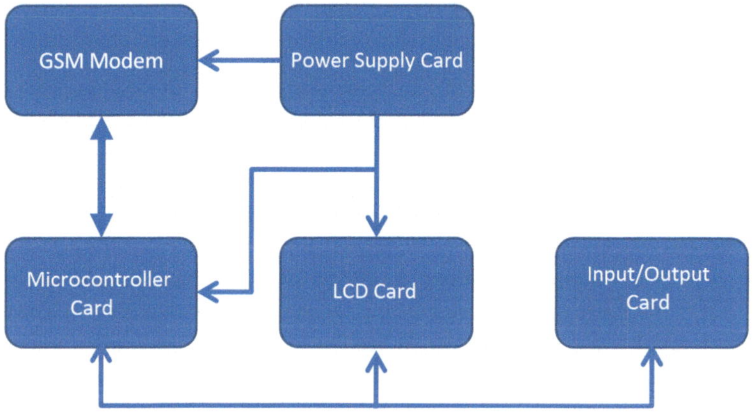

Fig. 3 Block diagram of the system

The system has also a mechanism to inform the user when the power fails via a text message through a GSM modem; thus users can take the necessary action.

The operation of the system and the block diagram of the system are given in Figs. 2 and 3, respectively.

Figure 4 shows a schematic diagram of the example of mini-refrigeraters in hotel rooms being monitored from a single place connected to the same network.

The product deficiencies including electrical faults in each refrigerator in the hotel rooms can be identified, procured, and priced without notice by the hotel customer.

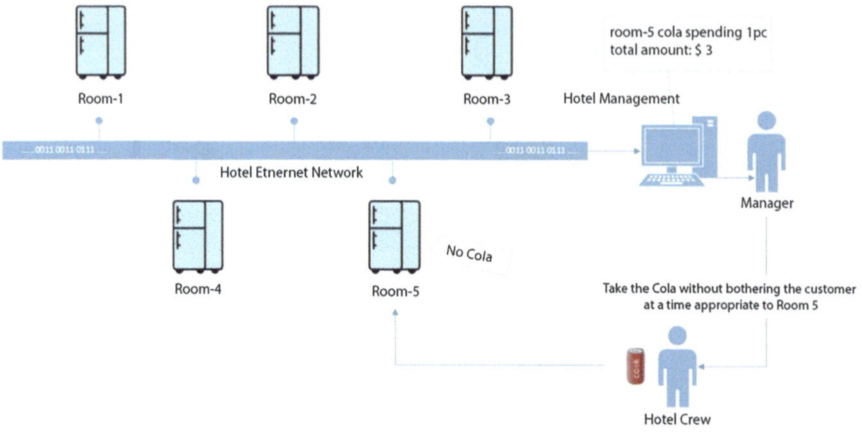

Fig. 4 The functioning of the system within a hotel

3.2 Working Principle and Algorithm

The project utilizes basic standard hardware components; a GSM module, and a micro-controller with I/O units connected to switches and sensors. The operation shows remote monitoring of the refrigerator (temperature and food and beverage contents and power failure) and generates alarms via text messages sent to the stored number in response to a call or message request or initiated by the system during power outages.

The software is a simple conditional looping using C, and standard AT commands are used for communication between the microcontroller and the modem. The user's phone is stored in the microcontroller. When the user makes a query by call or SMS, the latest status of the product information is sent to the GSM modem via RS232 port, and the information is transmitted to the user by the GSM modem as SMS. At the same time, freezer and cooler temperatures in the refrigerator are notified to the user. The information coming from the temperature sensors is converted to digital information by using ADC (analog-digital converter) ports of the microcontroller and sent to the user as SMS.

C language was used as a software language. The codes were compiled with the MPLABX C18 compiler and converted into a microcontroller. These systems serve some periodic and nonperiodic tasks. The flow diagram of the software is shown in Figs. 5 and 6.

Information can be received from the smart mini-refrigerater both via the WEB interface and GSM network. For this purpose, the mode selection switch on the front panel of the smart mini-refrigerater should be preselected. As shown in Fig. 5, if the GSM mode is selected, the microcontroller is waiting for the query. If a phone call or SMS is sent to the SIM number on a GSM modem with any mobile phone, the query is detected, and the presence/absence and number of products in

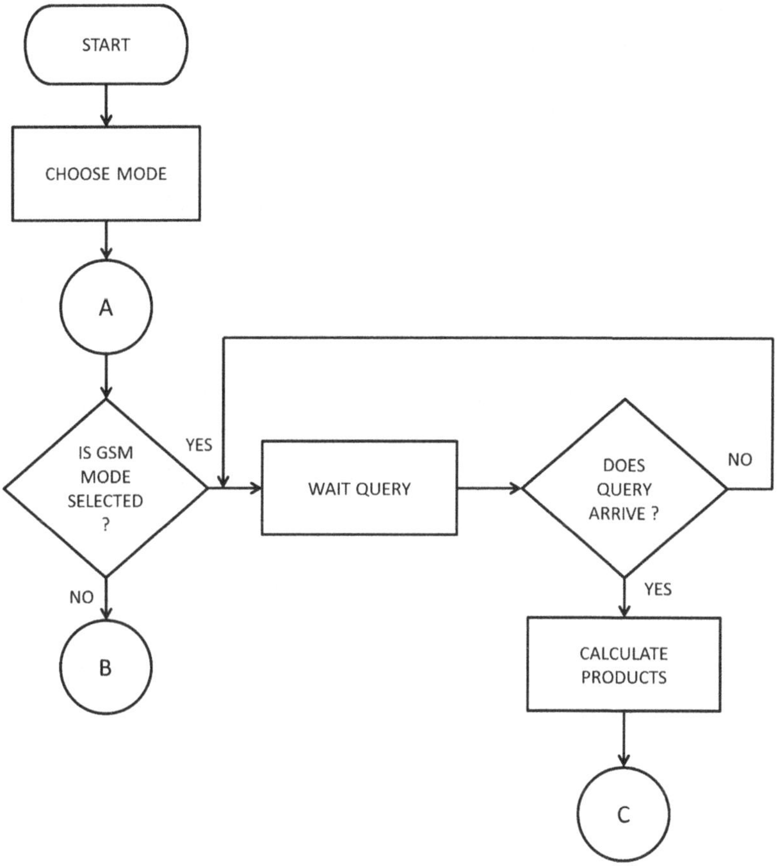

Fig. 5 Flow diagram of the software

the smart mini-refrigerater are calculated, and the information is transferred to a variable. Then, the SMS preparation process is started to inform the user.

Figure 6 shows a flow diagram showing how the SMS sending process takes place. AT codes are used by the GSM modem for sending SMS. These AT commands are sent to the GSM modem via RS232 serial communication port via microcontroller.

First, the number of the requesting phone is sent to the GSM modem. Then, the data is taken from the variable where the information is stored and the cooler and freezer temperatures of the smart refrigerator are read from the sensors by the microcontroller. All this information is converted to the appropriate format and then sent to the GSM modem. After the content of the SMS to be sent is completed, the SMS sending command is forwarded. This completes the process of sending SMS. The program starts waiting to receive the next query.

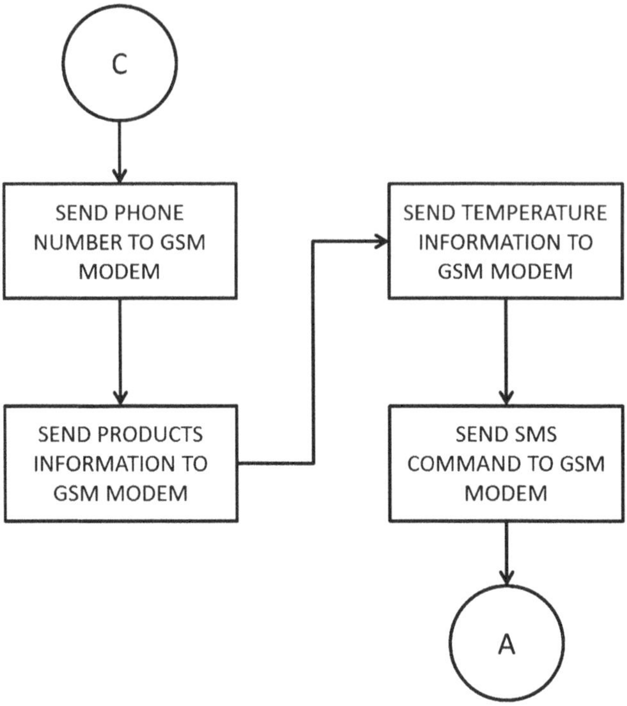

Fig. 6 Algorithm for sending text message subprogram

So far, we explained how the smart mini-refrigerater operates in GSM mode. In Fig. 7, the flow diagram of the operations performed in WEB mode is given.

In WEB mode, the IP address is shown on the LCD screen of the smart mini-refrigerater. When this IP address is typed into the address line in any web browser, the HTML page published by the microcontroller is displayed. On this page, the products in the smart mini-refrigerater can be followed instantly. As shown in Fig. 7, the products are calculated, and each product information is kept in separate variables and written to the XML file. When displaying the HTML page, the variables in this XML file are read and displayed instantly on the page.

Figure 8 shows a flowchart of notifying the user of the power failure by SMS. When the INT0 interrupt of the microcontroller is triggered, the interrupt codes are executed, and the mains power supply is interrupted or detected.

If the value of the signal received at the interrupt INT0 is 0, the mains power is interrupted, and an SMS is sent to inform the user. When the mains power is cut, the accumulator in the system enters and continues to operate the system. In this way, a smarter mini-refrigerater can be communicated for a while. If the value of the signal received at the INT0 interrupt is 1, the mains power is detected, and the user is notified by SMS.

Figure 9 shows the sample reports and status messages.

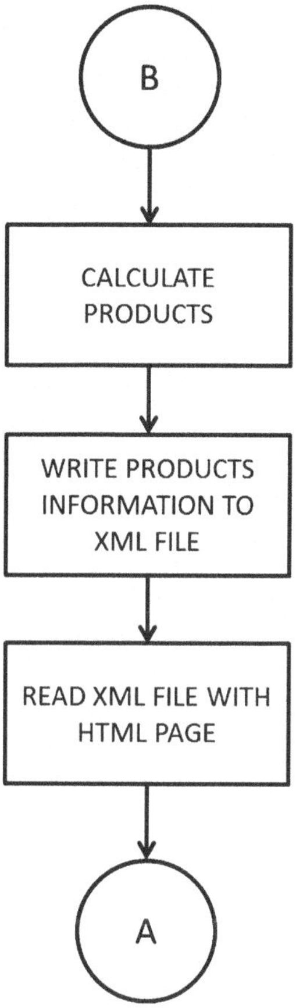

Fig. 7 The flow diagram of the WEB mode

4 Conclusion and Discussions

Due to the widespread use of intelligent systems, devices that provide user-machine interaction are becoming popular. This chapter describes a new refrigerator connected to both the GSM network and the Ethernet module. For this purpose, a prototype has been prepared, which enables the user to obtain information about the product quantities and temperature values from the remote refrigerator. In addition to allowing users to monitor their home devices from remote locations, an instant message informs them of all power outages in the system. New features may be added in the future as desired.

The smart mini fridge offered in this chapter has been tested by users in a laboratory environment for a period of 3–6 months. In addition, it was presented to the users' experience at the innovation-week fair, where innovative products are introduced every year in Turkey, both for the presentation of the project and for the demonstration of the operability of the system.

One of the objectives of the project was to integrate WEB-GSM technologies that make life easier in the mini-refrigeraters used in hotel rooms in the tourism

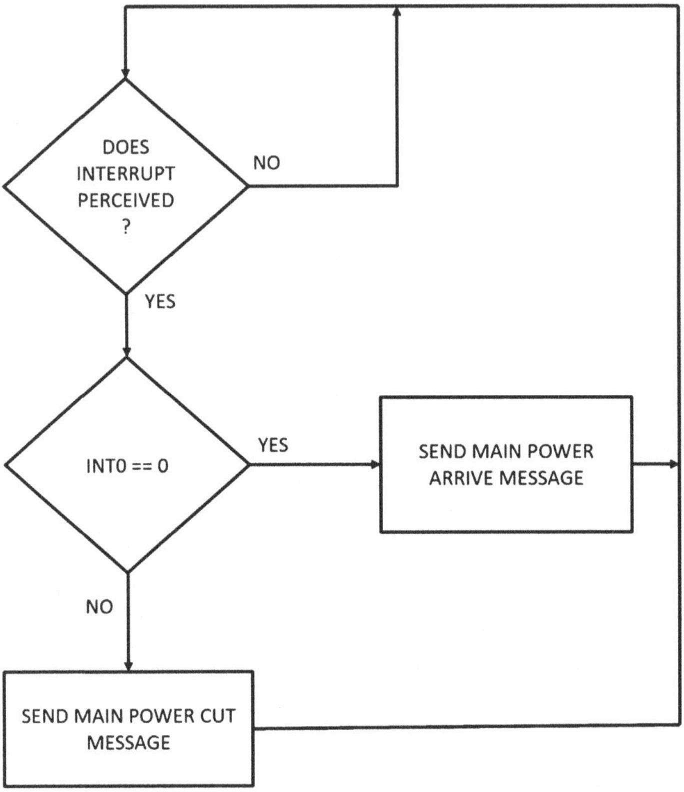

Fig. 8 Main power cut and return flow chart

sector. Monitoring the products that the customer uses in the mini-refrigeraters in the hotel rooms and the freshness of the products in the refrigerator will provide instant control, stock follow-up, automatic invoicing, more effective use of human resources in hotel management, and customer satisfaction.

Another target group of the project is the people with summer houses or farm/village/highland houses. While the project is being carried out, it is aimed to reach this target group by keeping them cheaper and widespread use. For example, this project is not only compatible with smartphone systems like Android, iOS, and Windows Phone, but all mobile phones and all computer systems.

Project output can have a wide market such as hotels, villa-type houses, etc. The project can be easily scaled up; this means that it can be used in larger cooling systems such as cold storage and can be developed with new technologies.

To conclude, it is seen that GSM and WEB technologies give fast and reliable results in smart home systems. These systems can provide significant advantages such as mobility, low cost, and electricity savings. It can provide similar benefits in daily life to people with disabilities. The project also aims to show that smart home systems do not require much expense in practice as commonly believed.

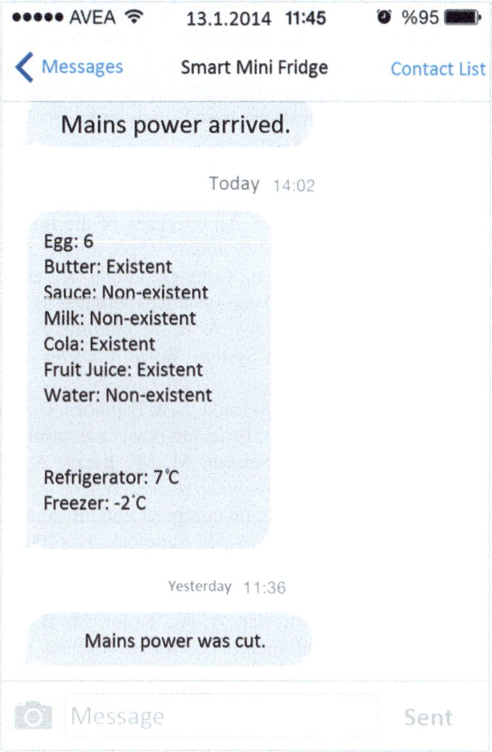

Fig. 9 The reports and status messages

The project can be easily scaled to large or small cooling systems such as cold storage or coolers in the cars. It can be integrated into smart home systems after improving with new technologies. In this context, it can interact with other smart devices.

Acknowledgments Our project is supported by TUBITAK by 2209/A-University Students' Domestic Research Projects Support Program. The device is patented under the National Patent No: TR 2014 05178 B.

References

1. da Silva, A. F., Ohta, R. L., dos Santos, M. N., & Binotto, A. P. D. (2016). A cloud-based architecture for the Interne-t of Things targeting industrial devices remote monitoring and control. *IFAC-PapersOnline, 49*(30), 108–113.
2. Wang, M., Zhang, G., Zhang, C., Zhang, J., & Li, C. (2013). *An IoT-based appliance control system for smart homes*, Fourth international conference on intelligent control and information processing, pp. 744–747.
3. Su, K., Cheng, H., Wang, H., & Lv, R. (2012). *Instant messaging application for the Internet of Things*, International conference on computer science and service system, pp. 166–169.

4. Cheuque, C., Baeza, F., Marquez, G., & Calderon, J. (2015). *Towards to responsive web services for smart home LED control with Raspberry Pi. A first approach*, 34th international conference of the chilean computer science society, pp. 1–4.
5. Alaa, M., Zaidan, A. A., Zaidan, B. B., Talal, M., & Kiah, M. L. M. (2017). A review of smart home applications based on Internet of Things. *Journal of Network and Computer Applications, 97*, 48–65.
6. Stojkoska, B. L. R., & Trivodaliev, K. V. (2017). A review of Internet of Things for smart home: Challenges and solutions. *Journal of Cleaner Production, 140*(3), 1454–1464.
7. Domingo, M. C. (2012). An overview of the Internet of Things for people with disabilities. *Journal of Network and Computer Applications, 35*(2), 584–596.
8. Welbourne, E., Battle, L., Cole, G., Gould, K., et al. (2009). Building the Internet of Things using RFID: The RFID eco system experience. *IEEE Internet Computing, 13*(3), 48–55.
9. Rajabzadeh, A., Manashty, A. R., & Jahromi, Z. F. (2010). A Mobile Application for Smart House Remote Control System. *World Academy of Science, Engineering and Technology, 62*, 80–86.
10. Chacon-Troya, D. P., Quezada, J., & Espinoza, C. (2017). *Development and implementation of a smart induction stove*, Brazilian power electronics conference (COBEP).
11. Rizvi, S., Sohail, I., Saleem, M. M., Irtaza, A., Zafar, M., & Syed, M. (2018). *A smart home appliances power management system for handicapped, elder and blind people*, 4th international conference on computer and information sciences, pp. 1–4.
12. Gürüler, H., İstanbullu, A., & Sarıçiçek, B. (2007). XML Biçimli, Belirsiz Değişkenlikteki Verilerin Aktarım ve Görselleştirmesine Mikro denetleyici Destekli Bir Çözüm Akademik Bilişim.
13. Yüksekkaya, B., Kayalar, A. A., Tosun, M. B., Ozcan, M. K., & Alkar, A. Z. (2006). A GSM, internet and speech controlled wireless interactive home automation system. *IEEE Transactions on Consumer Electronics, 52*(3), 837–843.
14. Shuhaiber, A., & Mashal, I. (2019). Understanding users' acceptance of smart homes. *Technology in Society, 58*, 1–9.
15. Martketsandmarkets, Smart home market by product (lighting control, security & access control, HVAC, entertainment, smart speaker, home healthcare, smart kitchen, home appliances, and smart furniture), software & services, and region - global forecast to 2024. https://www.marketsandmarkets.com/Market-Reports/smart-homes.
16. Abdulrazak, B. & Yared, R. (2015). *Prevent cooking risks in kitchen of elderly people: adaptable reasoning engine based on fuzzy logic for smart oven*, IEEE international conference on computer and information technology, pp. 2165–2172.
17. Kahan, T. H. (2018). Smart microwave oven with image classification and temperature recommendation algorithm. *International Journal of Electrical and Computer Engineering, 8*(6), 4239.
18. Abdullah, S. M. (2016). Design secured smart door lock based on Jaro Winkler algorithm. *Tikrit Journal of Pure Science, 21*(6).
19. Elnashar, E. A. (2018). *Self-cleaning technology of smart curtains and upholstery for theaters and hospitality*, International conference on materials research & development.
20. Ciabattoni, L., Freddi, A., Ippoliti, G., Marcantonio, M., Marchei, D., Monteriù, A., & Pirro, M. (2013). A smart lighting system for industrial and domestic use. In *Proceedings of the IEEE international conference on mechatronics (ICM)* (pp. 126–131).
21. Kumar, K., Sen, N., Azid, S. & Mehta, U. (2017). *A fuzzy decision in smart fire and home security system*, IEEE international symposium on robotics and intelligent sensors, pp. 93–98.
22. Anufrienko, A. (2019). *Appliances of smart TV as an IoT device for Industry 4.0*, IEEE 21st conference on business informatics (CBI), pp. 1–4.
23. Yamunathagam, D., Shobana, G., & Suguna, M. (2018). Smart care-predictive maintenance system in washing machine using IOT. *Journal of Advanced Research in Dynamical & Control Systems, 10*, 914–917.
24. Floarea, A. D., & Sgarciu, V. (2016). *Smart refrigerator*, International conference electronics, computers and artificial intelligence, pp. 19–24.

25. Wu, H. H., & Chuang, Y. T. (2017). *Low-cost smart refrigerator*, 1st International conference on edge computing, pp. 228–231.
26. Singh, D., & Jain, P. (2016). IoT based smart refrigerator system. *International Journal of Advanced Research in Electronics and Communication Engineering, 5*(7), 2080–2084.
27. Qiao, S., Zhu, H., Zheng, L., & Ding, J. (2017). *Intelligent refrigerator based on Internet of Things*, IEEE international conference on computational science and engineering and IEEE IC on embedded and ubiquitous computing, pp. 406–409.
28. Edward, M., Karyono, K., & Meidia, H. (2017). *Smart fridge design using NodeMCU and home server based on Raspberry Pi 3*, 4th international conference on new media studies, pp. 148–151.
29. Mahajan, M. P., Rohit, R. N., Vivek, P. P., & Rahul, D. D. (2017). Smart refrigerator using IoT. *International Journal of Latest Engineering Research and Applications, 2*(3), 86–91.
30. Nasir, H., Aziz, W. B. W., Ali, F., Kadir, K., & Khan, S. (2018). The Implementation of IoT based smart refrigerator system, 2nd international conference on smart sensors and application, pp. 48–52.
31. Zhongming, W., & Yanan, Y. (2018). *Design of an interactive smart refrigerator based on embedded system*, International conference on sensing, diagnostics, prognostics and control, pp. 589–592.
32. Luo, S., & Li, J. (2009). A smart fridge with an ability to enhance health and enable better nutrition. *International Journal of Multimedia and Ubiquitous Technology, 4*(2), 69–79.
33. Shweta, A. S. (2017). *Intelligent refrigerator using artificial intelligence*, 11th international conference on intelligent systems and control, pp. 464–468.

ChildPOPS: A Smart Child Pocket Monitoring and Protection System

Fatma Hussain, Issam Damaj, and Iyad Abu Doush

1 Introduction

Children safety is crucial for any parent, especially during work times. Parenting can be a challenging task; and with a busy modern lifestyle, this can be highly delicate while raising infants and toddlers. Today's families are composed of working couples; parents do work jointly to improve family's living standards. To that end, the constant and close monitoring of physical parameters of infants and their health conditions become an important matter. Since infants can't provide any feedback about their discomfort or health conditions, it is then safety-critical to tightly monitor infants. Furthermore, a famous parenting challenge is related to the danger of apparently life-threatening events (ALTE) or the sudden infant death syndrome (SIDS). According to the American SIDS institute, SIDS is defined as the sudden death of an infant younger than 1 year of age, which remains unexplained after a thorough investigation. According to The Lullaby Trust organization [1], there are approximately 300 unexplained infant deaths every year in the UK. With no doubt, the advancement in modern technology aids the development of smart

---

F. Hussain
Kuwait Aviation Services Company, Kuwait International Airport, Farwaniya, Kuwait
e-mail: F.hussain@kasco.com.kw

I. Damaj (✉)
Electrical and Computer Engineering Department, Beirut Arab University, Debbieh, Lebanon
e-mail: i.damaj@bau.edu.lb

I. Abu Doush
Computer Science and Information Systems Department, American University of Kuwait, Salmiya, Kuwait
e-mail: idoush@auk.edu.kw

M. M. Banat, S. Paiva (eds.), *Smart Technologies for Smart Cities*,
EAI/Springer Innovations in Communication and Computing,
https://doi.org/10.1007/978-3-030-39986-3_3

monitoring systems to help preventing and controlling children-related health risks by parents.

In this chapter, we present the development of a smart Child Pocket Monitoring and Protection System (ChildPOPS). ChildPOPS is concerned with ensuring infant's safety through real-time monitoring of multiple indicators. To present the system development, we adopt the IoT Development Model (IDM) of Damaj et al. [2]. Several development challenges are addressed including the need of redundant notifications due to the safety-critical nature of application, the adoption of adequate communication standards, and investigating feasible power options. The proposed system has five different subsystems that range from user bands, mobile devices, and web-enabled software solutions. This chapter includes discussing system safety, maintainability, usability, and ease of deployment, to name but a few. The main objectives of the presented investigation are as follows:

- Develop an effective-in-application smart gadget.
- Ensure child safety through automatic monitoring.
- Enable effective interventions.

This chapter is organized so that Sect. 2 presents related work. Section 3 presents the motivation and research objectives of developing ChildPOPS. Moreover, Section 4 details the system development, and Sect. 5 explains the main implementation stages. Section 6 presents a thorough analysis and evaluation of the developed systems at the component, integration, and system levels. Section 7 concludes the chapter and identifies pointers to future directions.

2 Related Work

The literature is rich with solutions proposed for monitoring infants using modern electronic devices. Ramesh et al. [3] proposed a smart baby cradle that enables monitoring by parents, plays music, talks to the baby, checks baby's temperature, and detects if the bed is wet. The proposed system uses a camera and sensors to implement the different tasks. Moreover, Lavner et al. [4] present automatic algorithm that uses machine learning to detect if the baby is crying by analyzing the audio signal. A convolutional neural network (CNN) is used to distinguish if the baby is crying from other sounds, such as opening a door, people talking, etc. Sharma and Mittal [5] investigate acoustic characteristics of the baby cry sound to identify the reason. The melody contour, Mel-frequency cepstral coefficients (MFCCs), and harmonics factors are analyzed to identify the type of infant cry. Seven different cry reasons are categorized and identified using the sound signal. Furthermore, Schiavone et al. [6] developed a micro-fabricated device with sensing capabilities that uses kinematic sensing unit and a gaze-tracker device worn on the infant head to monitor infants gaze and head movements. The system measures, records, and processes the collected data about infants and sends it to the parents using wireless communication. Zakaria et al. [7] propose a small device to monitor the body temperature of babies using wearable sensors. The device can alert parents when the baby's body temperature is higher than normal. The system can send the information to the parent's mobile phone which can help them remotely monitor the baby.

Zhu et al. summarize different solutions proposed for monitoring infants. Some solutions utilize CO_2 sensors which are placed around the infant to provide information about respiration [8, 9]. Other systems are reported in the literature that use sensors to detect the infant's temperature from the ear. The proposed system was challenged by the attained accuracy due to the location of sensors [10]. Moreover, other systems included textile electrodes that can be integrated in different positions of the infant cloth. In another system, a triaxial accelerometer is used to detect abnormal wave motion in a bathtub to prevent infants from drowning in home swim pools. The system can give warning to safeguards using a wireless communication receiver; this helps in preventing infant drowning in the bathtub [11].

Several smart monitoring systems exist in the market with the purpose of fulfilling parents' needs of kids' safety. Some monitoring systems are one-way auditory, while other aural monitors transfer sounds both ways which allow baby to listen to parents' voice. Other baby monitors are one-way visual monitors, while other visual monitors have two cameras where the baby can also see parents. Some monitors are wired, and others are mobile or wireless. Most aural and visual monitors have auxiliary features such as playing lullabies and colorful lights. With recent technologies, the development of baby monitors, nowadays they can take a baby's body temperature, baby movement, and other features which help parents to monitor their kids and ensure their safety. Some monitors come in the form of a temperature detecting teddy bear, while others alert the mother by notification sounds from the device or by vibration. The following are sample products that are like ChildPOPS in some respects (see Table 1 for a comparison of main features); indeed, ChildPOPS provides six essential features that are not necessarily available in totality in other systems:

- *Bellman Baby Monitor:* The system is a wireless device that monitors baby cry. Alerts include vibrations and an illuminating LED. The operating range of the product is up to 250 m, depending on obstacles [12].
- *Teddy the Guardian:* The system includes a teddy bear that enables monitoring the health of a child. When being cuddled, it simply starts tracking child's heart rate, level of oxygen in blood, and body temperature. The teddy bear sends the collected data wirelessly to a tablet, smartphone, or web application [13].
- *Mimo Baby:* The system is an organic cotton kimono fitted with noncontact machine washable sensors. Mimo Baby monitors infant's respiration, skin temperature, body position, and activity level. The system supports Wi-Fi and Bluetooth connectivity and can stream the baby's sound to parents' smartphone using an application [14].
- *The Heart Guardian:* The system is a wireless device that monitors baby's pulse and temperature to prevent sudden infant death syndrome SIDS. The system has child and parent subsystems and includes a speaker and a display [15].

Table 1 Comparison among different related products and ChildPOPS

Product	Wireless	Body temp.	Cry detection	Heart rate	Room temp.	Humidity	Oxygen levels	Activity level	Body position
ChildPOPS	✓	✓	✓	✓	✓	✓			
Bellman [12]	✓		✓						
Teddy [13]	✓	✓	✓	✓			✓		
Mimo Baby [14]	✓	✓						✓	✓
Heart Guardian [15]	✓	✓		✓					
Withings [16]			✓		✓	✓			
Owlet [17]		✓		✓			✓		✓

- *Withings Smart Baby Monitor:* The system allows parents to monitor their baby using mobile devices using a high-resolution video. Parents can speak with their baby, play lullabies, or turn on the multicolor night light. Parents can monitor noises, motion, temperature, and humidity in the baby's room [16].
- *Owlet Baby Monitor:* The system is mainly a smart sock that measures temperature, heart rate, and blood oxygen level. In addition, it tracks infant sleep and alerts parents if the infant rolls over onto their front side while asleep. It sends the data via Bluetooth to parent's smartphone. In addition, it has an iPhone app that lets parent keep track of nursing information, and parents can view their infant's health statistics via the Internet using a computer [17].

ChildPOPS is a typical smart device that runs within an IoT environment. To present the development of the system, from concept to prototyping and through various design decisions, we adopt the IDM presented in [18, 19]. The attraction of using such a model is its support of effective visualizations of design stages using a set that comprises a Concept Refinement Pyramid (CRP), Decision Trees (DTs), Architecture and Organization Diagram (AOD), Communication Interfaces (CIs), Use Case Diagrams (UCDs), a Realistic Constraint List (RCL), and a Menu of Analysis Metrics (MAMs). The variety of IDM models and the conciseness of representation are major advantages of adopting IDM.

The CRP is a pyramid model that highlights the system functions and technologies. The CRP comprises three levels; each should be described with short statements. At the top level, a title is added to briefly represent the intended system concept. The remaining levels aim at expanding the system concept into a set of statements that presents the design and main parts of implementation—issues related to professional ethics and socioeconomic factors can be included within the CRP.

The functional requirements of the system are captured using the DTs, UCDs, and AODs. Multi-leaf-node trees represent the DTs; it presents alternate options of design procedures or the choice of an implementation technology. A highlighted node presents the DT's final selection and the recommendation of the designer. The UCDs are traditional UML models to explain the interactions between the system functionalities and actors, such as using, managing, including, and extending. System events flow is also denoted using UCDs. The architecture and organization of the system are illustrated by the AOD. Furthermore, the AOD depicts the CIs of the different subsystems.

The nonfunctional requirements of the system are presented using the RCL and MAMs models. An RCL enables the reasoning of the realistic constraints governing the system development or its future expansions. The RCL aids the clear identification of constraints and permits foreseeing possible limitations in the system. In addition, MAMs is of two main parts, namely, general metrics and the application-specific metrics. Typical metrics that are common across IoT systems are listed under general metrics, while application-specific ones are tightly related to ChildPOPS.

3 Motivation and Objectives

Child monitoring systems have been dramatically developed recently and helped in controlling SIDS, ALTE, or accidents that might happen to a child. Nevertheless, recognizably they are still lacking real-time monitoring and reporting. ChildPOPS aims at addressing such problems to provide improved child monitoring. The purpose of this investigation is to develop a smart Child Pocket Monitoring and Protection System (ChildPOPS). ChildPOPS automatically monitors kids' health. The IoT Development Model (IDM) of Damaj et al. [2, 18, 19] is used to design, represent, and analyze the system through a set of submodels. The objectives of the presented investigation are as follows:

- Ensure child, mainly infant, security and health and safety.
- Monitor the ambient environment of an infant.
- Monitor the physiological parameter of an infant.
- Enable seamless communication between infants and parents.
- Make parents feel less stressed or anxious when leaving the child with the babysitter or alone in their cribs.
- Produce a product with a reasonable price and maximum benefit to be affordable to almost all people.
- Develop an effective-in-application smart gadget.
- Adopt an advanced lifestyle using modern and ubiquitous technologies.
- Support accurate monitoring.
- Enable effective interventions.

The main challenges that the proposed system addresses include supporting accurate biometric measurements, remote sensing, correct detection, and adequate intervention. The proposed system is supported by two mobile monitoring devices at the child and parent, or caregiver, sides, sensors, wireless communication nodes, control processors, Internet access, and a supporting server. ChildPOPS enables a ubiquitous access using tablets and smartphones through a mobile app or a web browser. This chapter includes studying the effectiveness of adding smart bands, at the child and parents' sides, to enable accurate monitoring. In addition, the system supports the interaction with a software system that provides assisted communication and intervention when needed. System testing includes multiple detection scenarios and the reliability of interactions. System testing, analysis, and evaluation confirms ChildPOPS effectiveness in application.

4 ChildPOPS Development

This section focuses on representing the design models, explaining the different design alternatives and choices that were considered for ChildPOPS, and an in-depth study of each design alternative is done in order to choose the most suitable one to implement. The advantages and disadvantages of different hardware components are explained to justify the appropriate choice based on the project requirements. At this point, an AOD is presented to depict the various system components and their interfaces. A Use Case Diagram (UCD) is provided to further clarify the chosen design. In addition, this section includes the RCL and MAMs diagram to articulate the nonfunctional requirements of ChildPOPS.

ChildPOPS uses two different wireless technologies, namely, ZigBee and Wi-Fi, and aims at reliable, portable, and flexible control. ChildPOPS is ubiquitous as it can be accessed from different devices, such as tablets, personal computers, and smartphones, and supported by a mobile application and an application server. The proposed system communicates with mobile networks using short message service (SMS). Multiple sensors are distributed around the infant to communicate environment- and body-related information. These sensors are connected to micro-controllers and wireless transceivers. Moreover, all the collected data is uploaded to a webserver database to be used later. ChildPOPS CRP diagram in Fig. 1 shows the main concept of the system, the supported functionalities, and the technologies used.

Different implementation technologies are considered for the processing units of ChildPOPS. Options include Field-Programmable Gate Array (FPGA) boards from Altera [22], data acquisition (DAQ) systems from National Instruments (NI) [23], Arduino-based components [24], and Phidgets interfacing boards [24]. Instead of being restricted to any predetermined hardware function, an FPGA allows to program product features and functions, adapt to new standards, and reconfigure hardware for specific applications even after the product has been installed in the field [25]. NI devices enable high-performance interfacing and control options. With the availability of small-scale DAQ systems, such as myRIO, NI devices combine the strengths of FPGAs and accurate interfacing.

The Arduino project provides a rich set of microcontrollers and their easy-to-integrate components. Arduino makes several different boards, each with different capabilities. Arduino Mega is rich in interfacing capabilities. Furthermore, Arduino Mini is the simplest and most popular board that contains everything needed to support the microcontroller. LilyPad Arduino is a wearable e-textile technology designed with large connecting pads. It can be sewn to fabric and connected to power supplies, sensors, and actuators with conductive thread. On the other hand, Phidgets are USB sensors and controllers that connect computers to the real world. Phidgets offer a wide range of language support, which includes an API for C#, C++, Java, Python, Visual Basic, Ruby, LabVIEW, Max/MSP, and more. Example

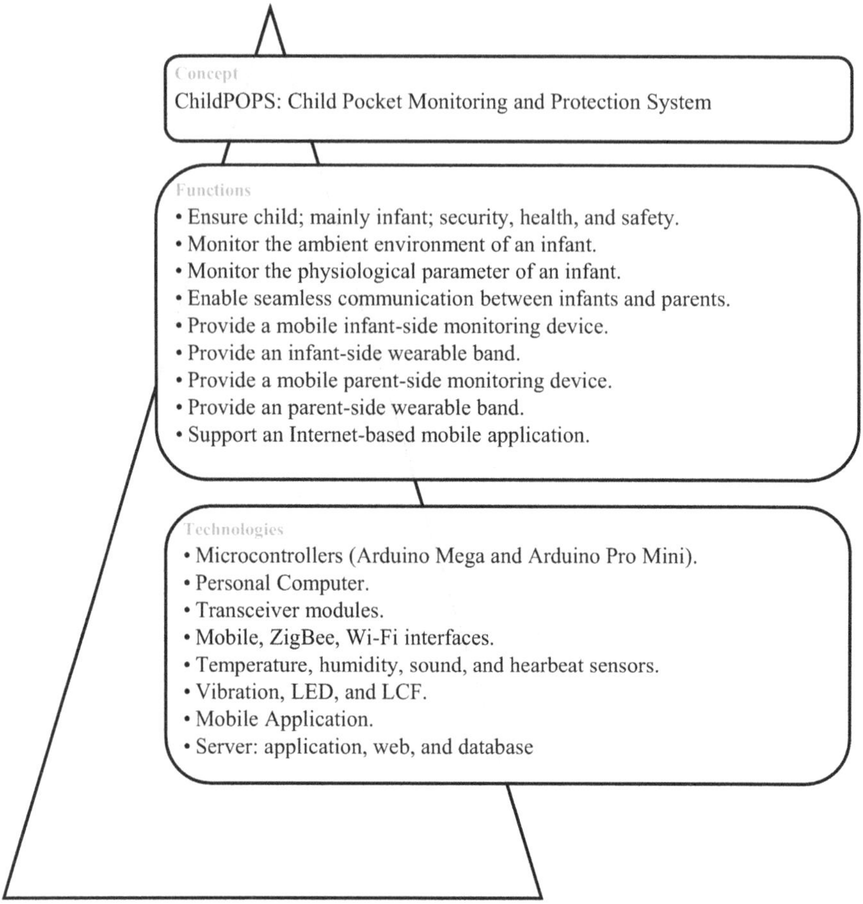

Fig. 1 Concept Refinement Pyramid (CRP) of ChildPOPS

codes for Phidgets, and for most products in several languages, are readily available and can easily be adapted for any project. Phidgets are computer-dependent, needing to be connected to a computer to run the application code.

The DT diagram shown in Fig. 2 is used to decide on the type of processing unit to use in ChildPOPS. Among the important factors that drive the decision are the size of the processing board, number of I/O bits, system extendibility, and ease of programming. All options, but the Arduino, suffered from their large board sizes that didn't fit well in the designed subsystems. The final decision is for using Arduino Mega for the mobile monitoring device, while using an Arduino Mini for the parent and infant bands.

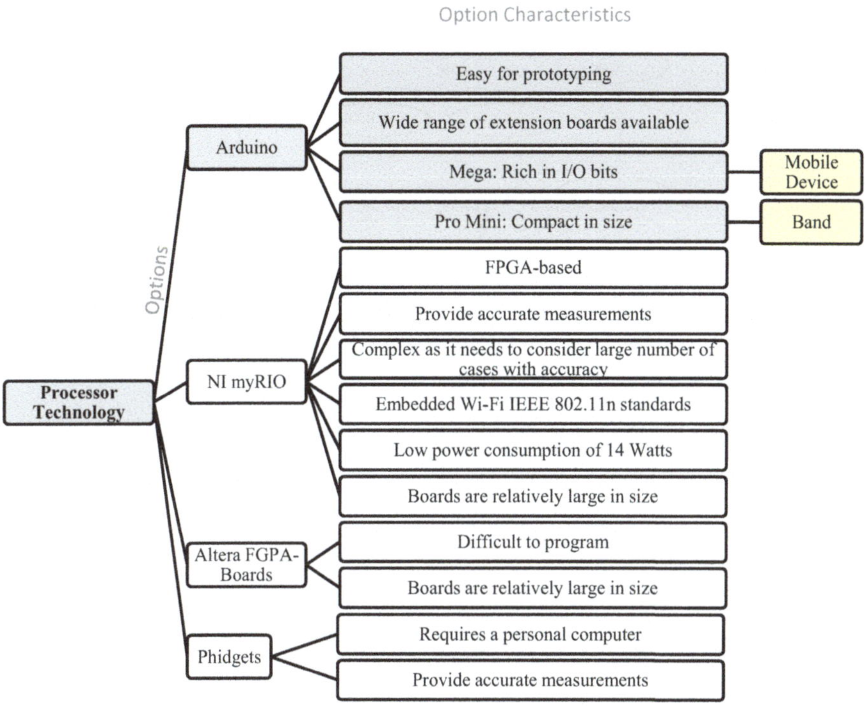

Fig. 2 ChildPOPS processor technology DT. The final decision branch is highlighted

The Use Case Diagram (UCD) of ChildPOPS is depicted in Fig. 3. The UCD captures the behavior of the system. The figure shows the parent's four options using ChildPOPS including a mobile application, a web interface, a band, and a mobile device. The devices on the parent's side enable multiple communication options including visual displays and vibrations. On the infant's side, the figure highlights the use of a band and a mobile device to sense the various parameters and update the system database through a server. In addition, both the band and the mobile device can connect to the server. Through the server, the parents can receive emails and SMS message to deliver alerts and information about the infant's status.

ChildPOPS architecture and organization employs a large variety of hardware and software subsystems. The hardware subsystem includes Arduino Mega and Pro Mini processors. In addition, ChildPOPS hardware includes bands, LCDs, and buzzers; vibration motors; LEDs; and sound, temperature, pulse, and humidity sensors. All devices are compatible with the Arduino family of microcontrollers. The software subsystem includes mobile and PC applications, a web server, and a supporting database. Figure 4 depicts the AOD of ChildPOPS.

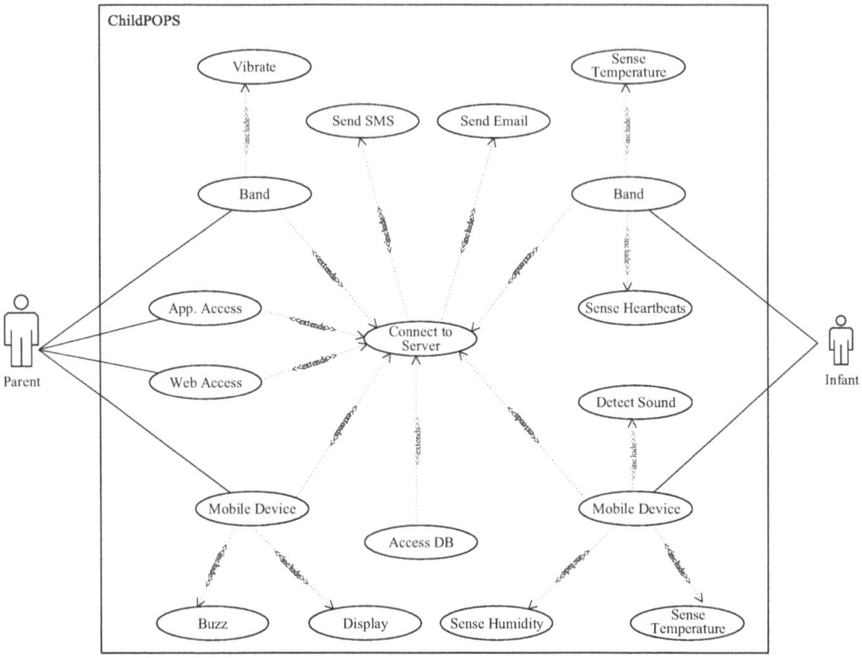

Fig. 3 ChildPOPS UCD

5 Implementation Aspects

The proposed system is incrementally implemented and tested on the component, system, and integration levels. The system implementation is divided into different parts, namely, the parents and infant bands and mobile devices and the software subsystems. The software subsystems comprise a web and application server that connects to an online database. Figure 5 shows the internal circuitry of the infant's mobile device. The components of the system include an Arduino Mega board, Wi-Fi shield, XBee shield, antenna, sound sensor, and humidity and temperature sensors. The infant's mobile device receives information from the infant's band through an RF transceiver and relays it to the server via its Wi-Fi connection. The collected data is for the sound, humidity, and temperature, in addition to the infant's band data that includes body temperature and heartbeat rate. In Fig. 6, the internal circuitry of the infant's band is shown. The infant's band is housed in a teddy bear with its sensors that can be attached to the body through flexible wiring.

The parent mobile device and band are shown in Figs. 7 and 8. The components used are an Arduino Mini Pro, XBee shield, antenna, vibration motor, and voltage regulator. The parent's band receives notifications using the RF transceiver from the mobile device—which receives data from the server via the Wi-Fi connection.

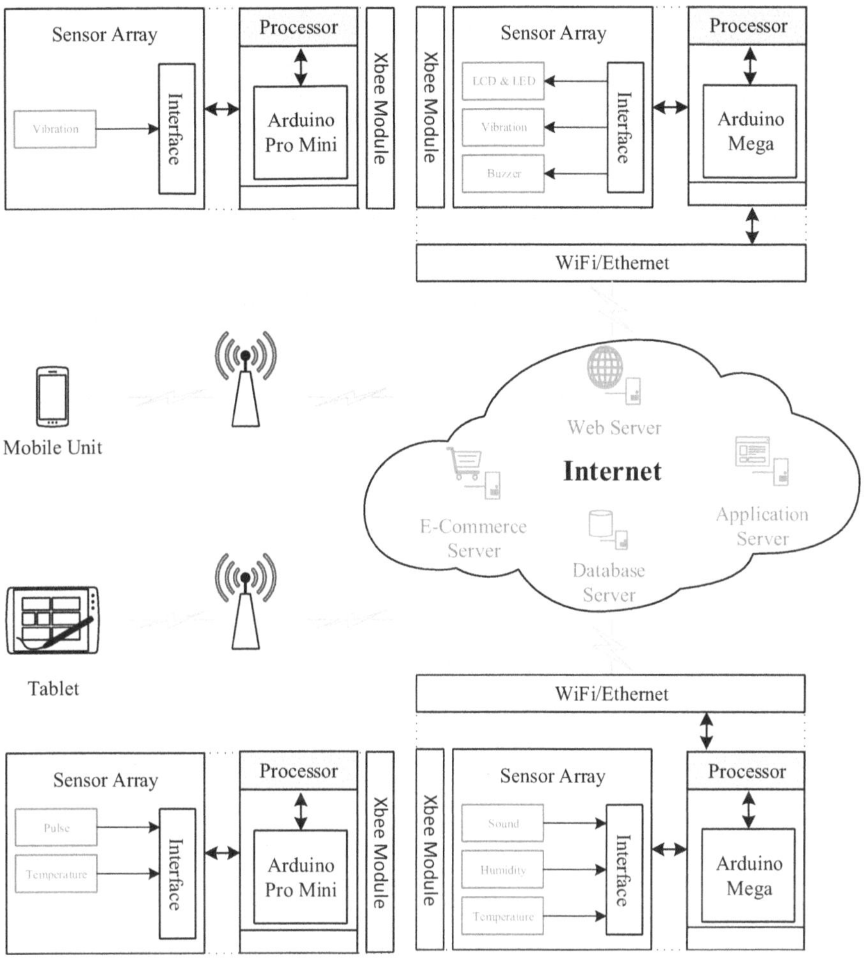

Fig. 4 ChildPOPS Architecture and Organization Diagram (AOD)

Figure 9 shows multiple measurements of sound in decibels (db), percentage relative humidity, and temperature in Celsius (°C).

Different software tools are used in the development of ChildPOPS. Arduino's Integrated Development Environment (IDE) is used to program the different microcontrollers. Microsoft Visual Studio is used to develop the PC-based interface. Furthermore, GoDaddy's online service is used to host the server and the database of ChildPOPS. Furthermore, JavaScript under MobiOne Studio is used to create an iOS application. The main function of the mobile application is to allow viewing the sensor readings at the infant's side (see Fig. 10). The mobile application is developed for an Android operating system.

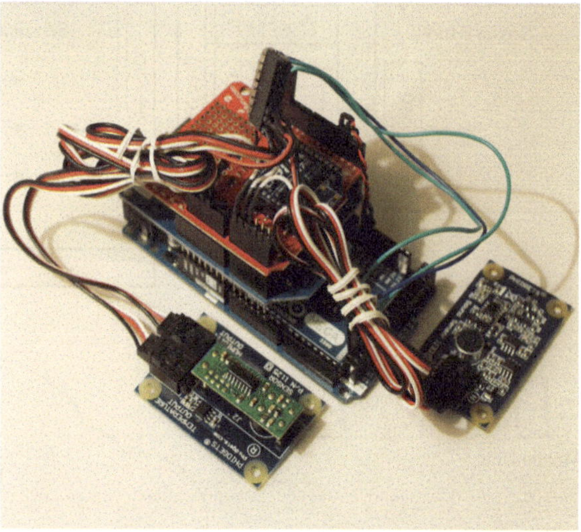

Fig. 5 Internal circuitry of the infant's mobile device

Fig. 6 Internal circuitry of the infant's band

6 Analysis and Evaluation

To analyze and evaluate ChildPOPS' effectiveness, the Menu of Analysis Metrics (MAMs) is presented in Fig. 11. The MAMs shows the nonfunctional requirements that are general and common in IoT systems. In addition, the figure shows a list of metrics that are specific to ChildPOPS.

ChildPOPS enjoys several general properties that comprise modernity, usability, portability, reliability, maintainability, and availability (see Fig. 11). ChildPOPS is

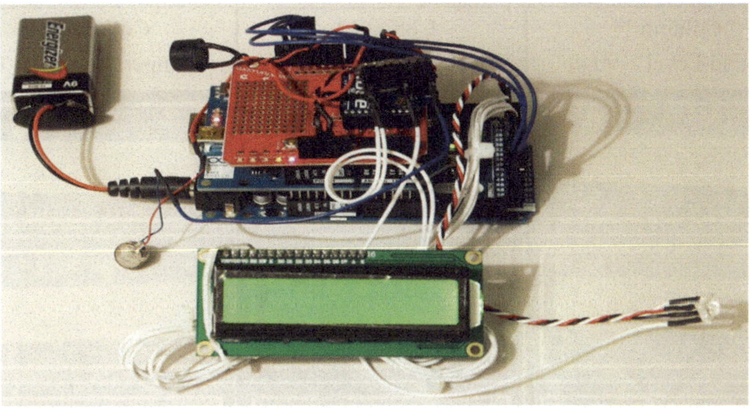

Fig. 7 Internal circuitry of the parent's mobile device

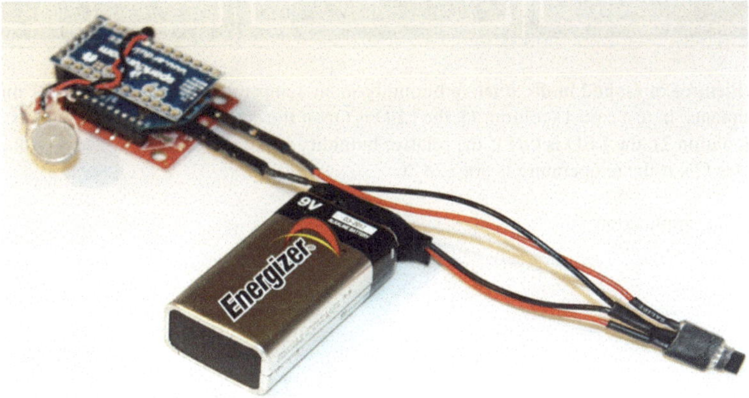

Fig. 8 Internal circuitry of the parent's band

modern in the sense that it supports features, such as a band, a mobile application, and the use of SMS and email messages. Moreover, the usability and learnability of ChildPOPS makes it a user-friendly system; naïve users can easily operate the system. In addition, ChildPOPS is portable as it can be used across different devices, such as laptops, PCs, or phones via mobile applications or web browsers. ChildPOPS exploits reliable web services that contribute to the real-time availability of the system. Indeed, the system has multiple notification options to ensure reliable operation in a safety-critical application; this feature is identified as one of the realistic constraints presented in Fig. 12. ChildPOPS is easy to maintain due to its simple assembly and the availability of components in the international and local markets.

In Fig. 11 several application-specific metrics are identified including communication range, response time, safety, efficiency, accuracy, and power consumption. ChildPOPS bracelets subsystems can communicate through RF wireless up to

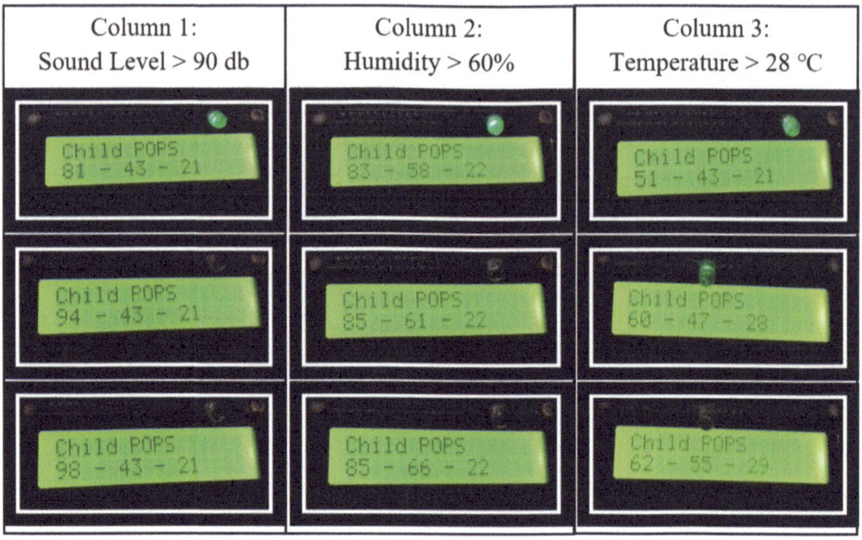

Fig. 9 Pictures of (sound in db, relative humidity in %, temperature in °C) readings of multiple measurements. In test case 1 (column 1), the LED is ON if the detected sound is not >90 db. In test case 2 (column 2), the LED is ON if the relative humidity is not >60%. In test case 3 (column 3), the LED is ON if the temperature is not >28 °C

Fig. 10 The application program of ChildPOPS

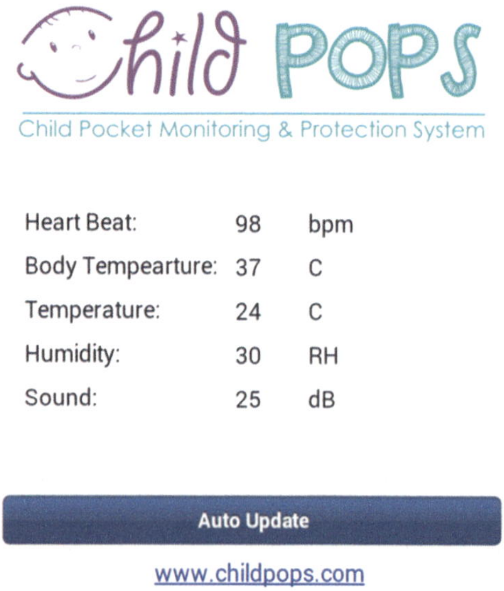

General	Application-Specific
• Modernity • Usability • Portability • Availability • Reliability • Maintainabiliy	• Communication range • Response time • Safety • Efficiency • Accuracy • Power Consumption

Fig. 11 Menu of Analysis Metrics (MAMs) of ChildPOPS

Redundancy — The system is for a safety-critical application that requires redundant notification options

- Use a variety of notification options (vibration, buzz, visual, sms, and email)

Wearable band — The infant's temperature and heartbeat sensors neeed to touch the body of the human subject

- Design a wearable band with flexible wiring and sensor interconnect
- The infant's band might suffer from wire disconnections due to movement; consider fixing the sensors as part of a wearable undershirt

Modern Lifestyle — Promote user satisfaction by providing a touch of modern lifestyle

- Benefit from modern technologies like mobile and web applications to provide a modern interface
- Use Wi-Fi communication due to wide availability

Wi-Fi signal strength — Weak Wi-Fi signal can cause disconnections in system commnication

- Consider using 4G shields

Power supply — The system prototype included using standard large-sized 12V batteries

- Consider the use of button battery cells
- Consider the use of rechargeable lithium batteries

Deployment proximity — The use of Wi-Fi limits the use of the system

- Consider using 4G shields

Fig. 12 Realistic Constraint List (RCL) of ChildPOPS

100 meters in range with both the infant-side and parent-side devices. Given the used RF and Wi-Fi connections, the system is suitable for deployment within apartments and can be expected to demonstrate the coverage of standard Wi-Fi router. Moreover, the current prototype of ChildPOPS uses 9 V batteries; based on extensive testing and continuous operation, the batteries supported the operation

of the system for several weeks. The shift to using button or rechargeable battery options is identified as possible future work (see Fig. 12). Based on thorough online testing, ChildPOPS is highly efficient in operation with satisfactory response times. The basic mobile application and web interface load quickly due to the light nature of the developed software. ChildPOPS is safe to use with no expected hazards due to the used low-voltage levels. Yet, ChildPOPS was tested on single-group human subjects that included one caregiver and one child. Furthermore, the training on using the system was straight forward; however, the delicacy in using the prototype will be eliminated upon the fabrication and packaging of ChildPOPS as a product. The system accurately detected the specified measurements during all testing phases. The realistic constraints of ChildPOPS, in the RCL diagram of Fig. 12, are specifications of the nonfunctional requirements of the system with several pointers to future work.

The following quantitative measurements capture the performance of the developed system:

- *RF range:* 100 meters for both the baby-side and parent-side devices.
- *Battery life:* 9 V batteries are used for the different devices; the batteries supplied power to the devices several weeks of operation, where the system is used regularly overnight and at least once during the daytime and for a couple of hours.
- *System speed:* Starting up the system before adding the Wi-Fi communication function takes less than 5 seconds, but after the Wi-Fi communication function is added, it takes up to 30 seconds for system initialization.
- *Response time:* The system responds to changes almost instantaneously—within less than 5 seconds using a landline DSL connection.

7 Conclusions

Child monitoring is a safety-critical application that can greatly benefit from modern technologies. Several challenging children death causes, such as SIDS, can be effectively controlled based on a solid infrastructure of mobile communication devices, software mobile applications, and the wealth of modern off-the-shelf hardware devices like the Arduino project. In this chapter, we have presented the design and implementation of ChildPOPS that can effectively enable the monitoring of infants' physiological parameters, besides other indicators, and communicate and intervene to maintain its adequacy. The design is presented using a modern IoT Development Model (IDM) that clearly captures the development decisions, system features, and analysis characteristics. The system is tested on the component, integration, and system levels and proved to be effective in application. Future work includes expanding the system features to cover measuring oxygen levels in the body, adding video monitoring, and adding communication interfaces, such as 4G shields. In addition, the system interoperability can be improved to function within a larger health informatics system. An essential future step is expanding the practical tests of the system to include a larger group of human subjects.

References

1. The Lullaby Trust. Statistics on SIDS, Retrieved on July 2019 from https://www.lullabytrust.org.uk/professionals/statistics-on-sids/.
2. Damaj, I., & Kasbah, S. (2019). Integrated mobile solutions in an internet-of-things development. In S. Paiva (Ed.), *Mobile solutions and their usefulness in everyday life* (pp. 3–31). Cham: Springer.
3. Ramesh, S., Hanidia Misbah, S. S., Bhavani, P., & Mamtha, H. L. (2019). A smart baby cradle. *Global Journal of Computer Science and Technology, 19*(1). https://computerresearch.org/index.php/computer/article/view/1815/1799.
4. Lavner, Y., Cohen, R., Ruinskiy, D., & IJzerman, H. (2016, November). Baby cry detection in domestic environment using deep learning. In *2016 IEEE international conference on the science of electrical engineering (ICSEE)* (pp. 1–5).
5. Sharma, S., & Mittal, V. K. (2017, October). A qualitative assessment of different sound types of an infant cry. In *2017 4th IEEE Uttar Pradesh section international conference on electrical, computer and electronics (UPCON)* (pp. 532–537).
6. Schiavone, G., Guglielmelli, E., Keller, F., Zollo, L., & Chersi, F. (2010, August). A wearable ergonomic gaze-tracker for infants. In *2010 annual international conference of the IEEE engineering in medicine and biology* (pp. 1283–1286).
7. Zakaria, N. A., Saleh, F. N. B. M., & Razak, M. A. A. (2018, July). IoT (Internet of Things) based infant body temperature monitoring. In *2018 2nd international conference on biosignal analysis, processing and systems (ICBAPS)* (pp. 148–153).
8. Cao, H., Hsu, L. C., Ativanichayaphong, T., Sin, J., Stephanou, H. E., & Chiao, J. C. (2007, March). An infant monitoring system using CO/sub 2/sensors. In *In 2007 IEEE international conference on RFID* (pp. 134–140).
9. Zhen, C., Xia, Z., Long, L., & Pu, Y. (2014). Accuracy of infrared ear thermometry in children: A meta-analysis and systematic review. *Clinical Pediatrics, 53*(12), 1158–1165.
10. Chen, W., Bambang Oetomo, S., Feijs, L., Bouwstra, S., Ayoola, I., & Dols, S. (2010). Design of an integrated sensor platform for vital sign monitoring of newborn infants at neonatal intensive care units. *Journal of Healthcare Engineering, 1*(4), 535–554.
11. Nishida, Y., Hiratsuka, K., & Mizoguchi, H. (2007, Oct, 28–31). Prototype of infant drowning prevention system at home with wireless accelerometer. In *Proceedings of the IEEE SENSORS 2007 conference* (pp. 1209–1212). Atlanta.
12. Bellman Baby Monitor. (2019). Retrieved on July 2019 from https://bellman.com/en/for-professionals/product-support/visit-notification-system/visit-transmitters-for-europe/baby-monitor/
13. Teddy the Guardian. (2019). Retrieved on July 2019 from http://teddytheguardian.com
14. The Mimo Baby Monitor (2019). Retrieved on July 2019 from https://www.mimobaby.com/
15. Claud, J., Carrillo, R., Wylie, J., Jensen, K., & Tchang, T. (2016). Sudden infant death syndrome heart rate and temperature monitor with audible and visual alarms. Retrieved on July 2019 from http://www.jason-skinner.com/wp-content/uploads/2016/08/SIDS.pdf.
16. Withings Smart Baby Monitor. (2019). Retrieved on July 2019 from https://www.withings.com/us/en/health-monitors.
17. Owlet Baby Monitors. (2019). Retrieved on July 2019 from https://www.withings.com/us/en/health-monitors.
18. Kandil, M., El-Attar, F., Al-Baghdadi, R., & Damaj, I. (2020). AmIE: An ambient intelligent environment for blind and visually impaired people. In S. Paiva (Ed.), *Technological trends in improved mobility of the visually impaired* (pp. 207–236). Cham: EAI/Springer Innovations in Communication and Computing. Springer.
19. Kandil, M., El-Attar, F., Al-Baghdadi, R., & Damaj, I. (2019, March 26–27). AmIE: An ambient intelligent environment for assisted living. In *2nd engineering innovations in healthcare international conference* (pp. 1–6). Dubai: IEEE.
20. Altera. (2019). Retrieved on May 2019 from https://www.altera.com.

21. National Instruments. (2017). Retrieved on September 2017 from https:/ http://www.ni.com.
22. Arduino. (2019). Retrieved on May 2019 from https://www.arduino.cc.
23. Phidgets. (2017). Retrieved on September 2017 from https://www.phidgets.com.
24. Damaj, I., Hawkins, J., & Abdallah, A. E. (2003, July). Mapping high-level algorithms onto massively parallel reconfigurable hardware. In *IEEE international conference of computer systems and applications* (pp. 14–22).
25. Qatan, F. M., & Damaj, I. W. (2012). *High-speed KATAN ciphers on-a-chip*, 2012 international conference on computer systems and industrial informatics, Sharjah, pp. 1–6.

Signal Processing-Based Automated Fault Detection Methods for Smart Grids

Gökay Bayrak and Alper Yilmaz

1 Introduction

Providing a reliable and stable power system is an essential subject to integrate a distributed generator (DG) into the grid. Smart energy management and protection system are also required for microgrids to prevent the malfunctions and power disturbances with a continuous grid operation. The faults in a power system affect the power quality, and power system will be out of the stability region in any faults [1]. Some power disturbances are voltage sags, voltage swells, and islanding in a power system [38], and they are the most common faults in a microgrid. Thus, energy management and fault protection system are essential for a safety grid connection. Figure 1 shows a photovoltaic (PV)-based DG connected to the grid, and in this DG, unintentional power disturbances can cause serious damages for both the DG and the operator. Thus, detection of the faults in a DG in time is a significant issue for a power system.

A microgrid system is out of reference values when power quality disturbances are occurring. Thus, a DG system must have a smart protection system to prevent the whole system in abnormal operating conditions [2]. It will not be possible to protect the microgrid system when the grid is not disconnected from the DG. Voltage sag/swell and islanding are significant fault conditions, and DG system and the grid are directly damaged in these faults. The grid loses the synchronization in these faults, and nominal operating conditions are not possible in the DG. The power system is unstable and not reliable because of losing the desired Grid Code values with the fault. Fault detection in a DG system, and selecting proper

G. Bayrak (✉) · A. Yilmaz
Bursa Technical University, Faculty of Engineering and Natural Sciences, Department of Electrical and Electronics Engineering, Bursa, Turkey
e-mail: gokay.bayrak@btu.edu.tr; alper.yilmaz@btu.edu.tr

M. M. Banat, S. Paiva (eds.), *Smart Technologies for Smart Cities*,
EAI/Springer Innovations in Communication and Computing,
https://doi.org/10.1007/978-3-030-39986-3_4

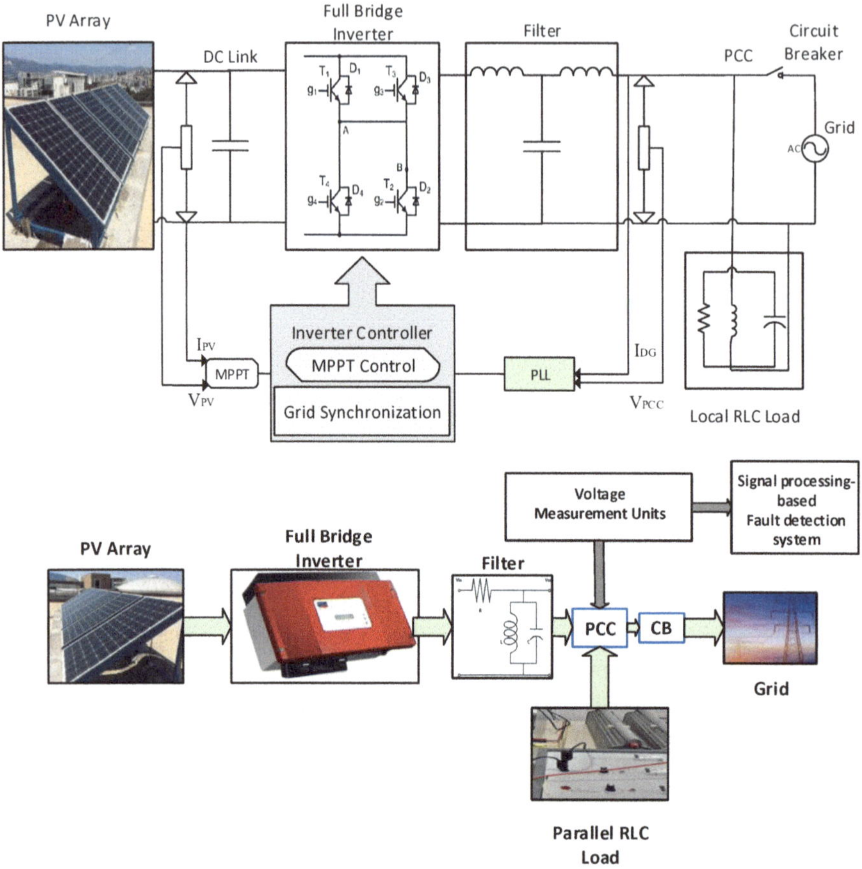

Fig. 1 A general schematic of a PV-based DG

protection schema for the microgrid, is a significant issue in the literature. The intelligent/smart fault detection methods are recently used in the literature rather than the conventional fault detection methods. Figure 2 shows conventional and intelligent fault detection methods in DGs. The intelligent fault detection methods are studied, and there has been an increase recently in the literature in studies on fault and power quality issues in power systems. On the other hand, there have been a limited number of studies using signal processing-based methods that automatically detected the faults in real time. In this study, for the first time, the structure of the undecimated wavelet transform (UWT)-based event recognition system developed to detect any faults for a microgrid in real time.

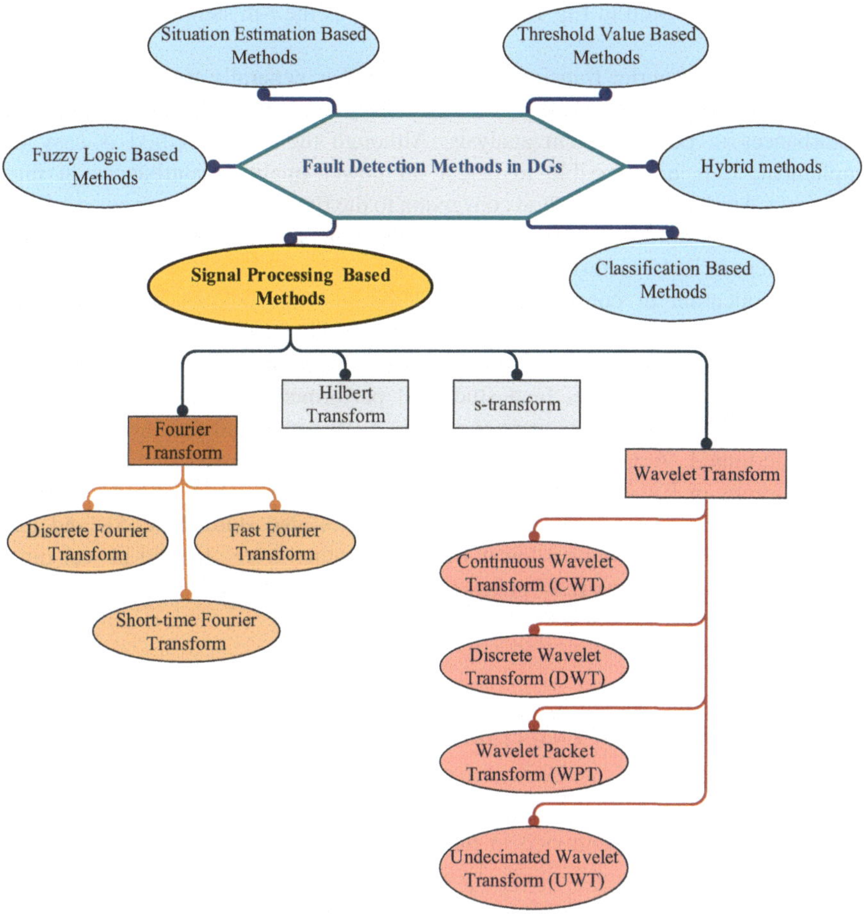

Fig. 2 Fault detection methods used in microgrids

1.1 Conventional Fault Detection Methods

The effective value transformation (RMS) is a fundamental fault detection method in international standards using a voltage signal to detect disturbances in a microgrid. The effective value conversion to the instantaneous voltage signal is implemented with the help of the following equation:

$$U_{\mathrm{RMS}} = \sqrt{\frac{1}{M}\sum\nolimits_{i=1}^{M} u(i)^2} \quad (1)$$

M is the window width in Eq. 1, and $u(i)$ represents the *i*-th sample of the voltage waveform.

Recently, fast Fourier transform (FFT) method is generally used for harmonic analysis, and the RMS method is used for detecting short- and long-time voltage disturbances in power system analysis. Although the RMS method is easy to implement, it is not possible to detect all power quality disturbances in this technique. Applying RMS method conversion to the fundamental frequencies, phase angle, and the harmonics of the signals are uncertain. Thus, the defined accuracy is not sufficient in RMS method.

The application of conventional methods is highly practical. Besides, difficulties in selecting threshold values, false detection conditions, noise effects in measurement, and uncertain fault conditions in several operating conditions make these methods less reliable. In a study [3] performed using a threshold-based method, threshold-based tests were applied to a PV panel, and the faults in panels were determined. A grid-connected PV system is investigated in [4] to detect the distortions in PV panels by comparing measured data with defined electrical threshold values. The island mode of operation is also studied in [5], and the upper/lower voltage and upper/lower frequency values were checked in real time with the help of the LabVIEW program, and the fault was detected according to these values.

1.2 Intelligent Fault Detection Methods

The conventional fault detection techniques are easy to implement, but they have several disadvantages mentioned in the paper. Recently, signal processing-based intelligent fault detection methods have been proposed in the literature. These methods use different procedures, including field transformation, fuzzy rules, event estimation, and classification processes. These intelligent techniques are reliable than conventional methods and provide detecting any faults in a microgrid.

Using the fuzzy logic method in intelligent methods makes a major contribution to conventional methods about providing the reliability and stability of the microgrid system [6, 7]. Determined data from predictions or real observers are compared in situation estimation methods to obtain the current state of the microgrid system [8, 9]. The faults and power quality disturbances in the electricity energy substructure are also investigated by using different classification methods in the literature. Artificial neural network-based classifiers are an improved method inspired primarily by cells called neurons in the human brain and are now becoming popular again with a significant increase in the processing capacities of computers. In the studies performed, neural networks were trained using real-time data, data based on mathematical equations, and data obtained as a result of the simulation. Then, the fault detection is achieved by using the output of the trained neural network [10, 11]. Many pattern recognition problems can be solved by support vector machines (SVM) method, and using kernel function that is used especially in multi-class pattern recognition problems helps to accomplish a successful classification [12, 13]. Determining the fault location in cables and transmission lines with a model-based algorithm by using SVM method is proposed in [14, 15]. DG systems generally use raw sensor data, and they are nonlinear systems. Thus, intelligent methods are required to detect any faults in a DG system. Using feature extraction methods for detecting the power disturbances allows better error classification than other methods [16, 17].

The frequency and time components of a signal are obtained in field transform methods, and using the frequency component provides a significant contribution to these methods rather than conventional methods [37]. WT and the FT constitute the field transform-based methods. Field transform methods generally use FT and WT techniques to detect the power quality disturbances in a DG.

WT method allows varying the window size and provides analyzing the signals in the time domain. FT only uses frequency domain analysis. Thus, WT has an advantage from FT by including time-domain analysis. Therefore, power quality disturbance signals consist of several frequency components, so the WT method presents a more reliable fault detection system comparing to other signal processing-based methods.

A voltage sag/swell event detection is investigated in [18], and Daubechies mother wavelets are used in the WT method to detect these events properly. The proposed WT-based method is also compared to the FFT method by using the PSCAD program, and the performance of the presented method is tested under different fault conditions. WT method is performed in wind turbines, and five levels of discrete wavelet transform (DWT) are tested to obtain coefficients [19]. A DG system consisting of PV panels and connected to the grid in [20] is investigated, and the current control embedded in the PV inverter and the WT technique was proposed for detecting the faults. ATP/EMTP software is used in [21] to realize the obtained results from the simulated model, and WT is tested and verified with several power quality disturbances. In one of the studies performed, filter banks including Daubechies wavelet family were used to propose a method of thresholding power distribution system signals determining energy, entropy, and harmonics by DWT decomposition [22]. Voltage change in short time, transients, and harmonic signals were studied in [22] by evaluating determined mathematical models. The results show that the WT algorithm has a successful fault detection capability in real-time signals. Wavelet packet transform (WPT) is also used for fault detection in DGs, and one of these studies [23] classification of three-level WT is performed in a DG to detect the power quality signals on time.

2 Detection of Power Quality Events with Fourier Transform

The WT method, laid down by A. Haar [36], which is frequently used in pattern recognition and power system fault detection applications in recent years, is a signal processing method capable of processing data in several resolutions and scales.

In contrast to the FT method, the WT method evaluates the event signal in a scale-frequency framework, which also includes time information. Besides, the fixed window size applied to the signal in the short-term Fourier transform (STFT) method becomes variable in the WT method. Naturally, in the analysis made with STFT, the window size remains constant for all frequency values; in spite of this, WT presents a flexible analysis. Power quality disturbance signals may include both high-frequency and low-frequency disturbance signals. Because of the different frequency signal components contained in the power quality disturbance signals, the WT method is preferred over the FT and STFT method.

In this section, FT and STFT methods are briefly mentioned for a better understanding of WT. Then, the wavelet theory, wavelet types, and WT methods are discussed. Finally, the power quality events occurring in the power system are analyzed by WT methods in MATLAB/Wavelet Toolbox environment.

2.1 Fourier Transform-Based Methods

The FT method, which is the most widely used signal processing technique established by Joseph Fourier, separates the signal into sine waveforms at different frequencies and provides a relationship between the time-frequency regions of the signal. That is, the time-based signal in the frequency domain corresponds to the FT. The FT formula is given in Eq. (2):

$$F(\omega) = \int_{-\infty}^{\infty} f(t) e^{-j\omega t} dt, \quad \omega \in (-\infty, \infty) \tag{2}$$

FT results, which provide a general approach to signal signaling, do not allow the local examination of frequency components. For this reason, the FT method gives positive results in stationary and continuous signals such as harmonic generation, DC injection, and noise generation, but it cannot be used because of the change in time and sudden changes in short-term voltage changes and transient events.

The problem of not being able to examine the signals with variable properties by the FT method was solved by windowing method, which allows the first signal used by Dennis Gabor (1946) to be divided into smaller components. In this transformation called the STFT, the signal is divided into short-term signals by windowing, and FT is calculated for each window. STFT formula is given in Eq. (3):

$$F(\omega) = \int_{-\infty}^{\infty} f(t) \omega (t - \tau) e^{-j\omega t} dt, \quad \omega \in (-\infty, \infty) \tag{3}$$

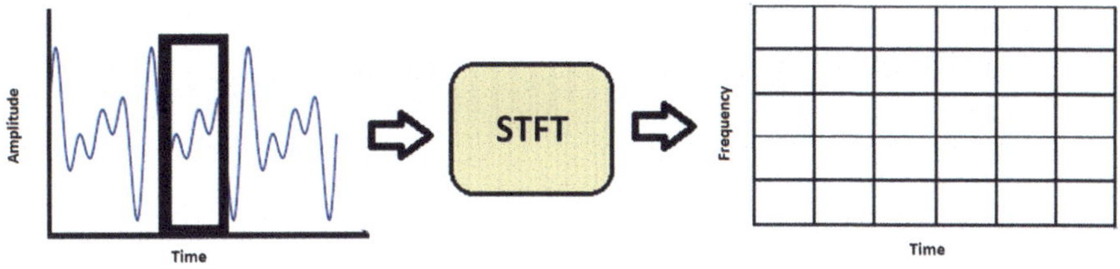

Fig. 3 Short-term Fourier transform

Here the signal $f(t)$ maps the time-frequency plane by multiplying the $(t-\tau)$ sliding window function selected by the user for analysis. The window functions commonly used in the analysis of power quality disturbances are Bartlett, Hanning, Kaiser, rectangular, and triangular window functions.

The STFT establishes a complete relationship between time and frequency, as shown in Fig. 3, and allows the frequency at which time the signal is generated, depending on the windowing size. The window size used is fixed, and the frequency resolution of the signal increases as the window size increases, while the time resolution increases as the window size decreases. With the STFT method, it is not possible to examine both the high- and low-frequency components of a signal at the same time [26]. This condition is the biggest disadvantage of the STFT method.

The over-/underfrequency change event is discussed to give an example of fault detection by the STFT method. The STFT-based method is applied to the voltage signal in real time at the point of PCC in microgrids. The over- and undervoltage fault is the deviation of the voltage from the nominal value in the DGs, resulting from overloaded motors, overloading, and faults along the line.

For the detection of over-/underfrequency fault by STFT analysis, Hanning function was selected as the windowing function, and analysis was performed for two different windowing lengths, $L = 32$ and $L = 64$. For analysis, the grid frequency is 50 Hz, and the sampling frequency is 1 kHz. In Fig. 4, a sudden frequency drop occurred in the voltage signal acquired with measurement board at 500 ms, and as a result, the underfrequency limit value was exceeded. With this decrease in frequency value, the new frequency value was 47.5 Hz.

3 Detection of Power Quality Events with Wavelet Transform

3.1 Wavelet Transform (WT) Method

The WT method, which has been frequently used in pattern recognition applications in recent years, was introduced by A. Haar [36]. WT method use data are consisting of several resolutions and scales. Although the other signal processing-based

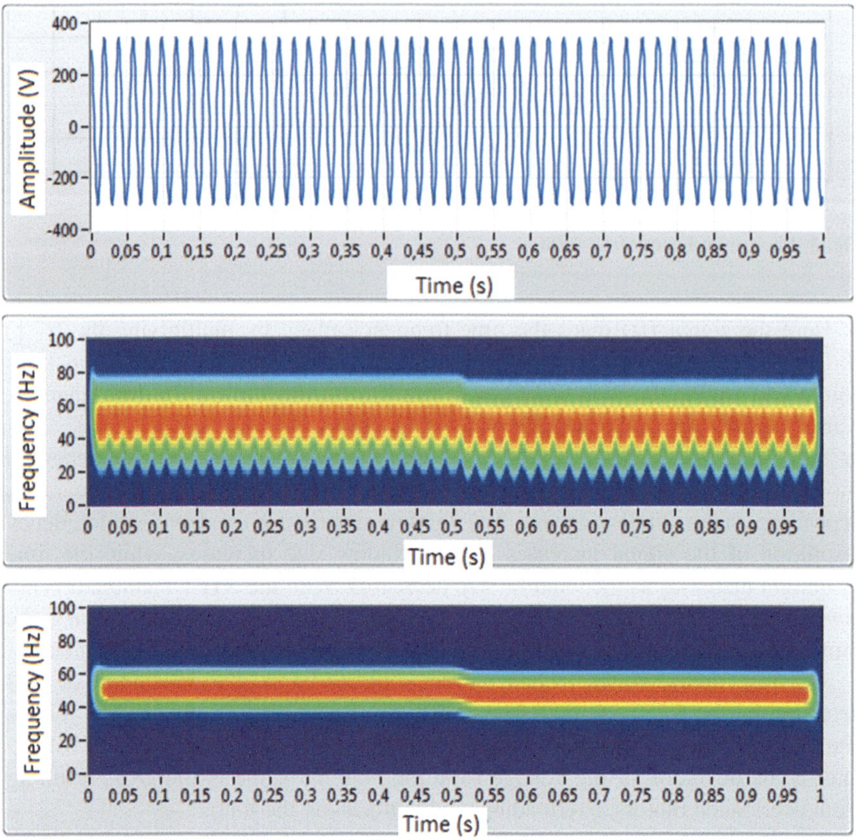

Fig. 4 STFT analysis of underfrequency fault

methods are not capable of detecting sudden changes and discontinuities in the processed signal, the WT method can determine these events.

The time data is not available in the WT method because the event signal is processed in the frequency and amplitude zone. Thus, WT is successful by detecting stationary signals like harmonics, but it is not sufficient to detect transients and short-term voltage changes. Besides, STFT detects nonstationary signals with high accuracy, but it is not possible to simultaneously analyze low- and high-frequency components of the signal because windowing is a constant scale in STFT [24, 25]. In the WT method, as shown in Fig. 5, unlike these two methods, the signal is analyzed on the time-scale axis to obtain a more flexible representation. Employing a flexible time-frequency representation, WT is applied to the signal. WT provides extending the transient and low-frequency events in a wide window scale. This allows the use of a narrow window to open high-frequency prone components.

The Haar mother wavelet, defined by Alfred Haar in 1909, is the first wavelet function. The Haar wavelet was first used by P. Levy to model the random motion

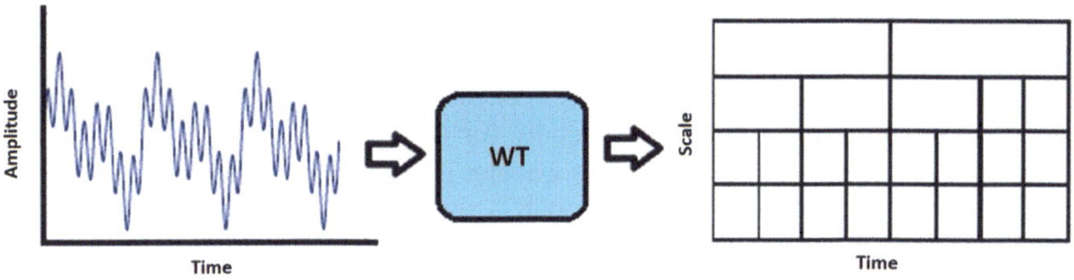

Fig. 5 Wavelet transform method

of particles instead of the FT method. The definition of the WT method was made by A. Grossman and Morlet, who worked at the Center for Theoretical Physics in Marseille, and the mathematical basis of the WT method was established [27]. In 1989, Stephane Mallat and Ingrid Daubechies formed the mathematical basis of discrete signal transformation. The multi-resolution decomposition method used to separate a signal into more than two sub-frequency values using WT and high- and low-pass filter by Mallat is still one of the most widely used methods. Subsequently, R. Coifman and V. Wickerhauser developed the WT, which is the continuation of multi-resolution separation [28].

Wavelet is defined as the scalable window style in the WT method, and it is used instead of the window concept used in the STFT method. Unlike the sinus sign in the FT method, the main wavelets are finite, irregular, and asymmetrical. The window function transforms into the main wavelet function ($\psi(x)$) in WT method, and it can be shifted and scaled during the signal analysis. The wavelet function can give clearer information about the signal if it is created for a short time in high-frequency signal analysis and for a long time in low-frequency values. To be called a wavelet function, the integral of the function must be zero, and the integral of the square of the same function must be one [29]:

$$\int_{-\infty}^{\infty} \psi(t)dt = 0 \tag{4}$$

$$\int_{-\infty}^{\infty} \psi^2(t)dt = 1 \tag{5}$$

The mathematical expression of the Haar wavelet is given in Eq. (6):

$$\psi_{\text{Haar}}(t) = \begin{cases} {}^{-1}\!/\!_{\sqrt{2}}, & -1 < x \le 0 \\ {}^{1}\!/\!_{\sqrt{2}}, & 0 < x \le 1 \\ 0, & \text{other conditions} \end{cases} \tag{6}$$

The Haar wavelet, the simplest function that satisfies mathematical expression conditions, is also a special definition of the Daubechies wavelet (dbN). This wavelet type found by Ingrid Daubechies is the most common type for power quality disturbances and fault detection, especially for DWT applications. The larger the wavelet number, the more frequent the wavelet is seen on the time axis, making it easier to detect high-frequency events, such as transient regime events. The wavelet functions commonly used in applications are given in Fig. 6.

Figure 7 shows the time-frequency representation of the wavelets used for WT. Here the parameters s and b correspond to the scale and displacement amounts, respectively. Increasing the amount of scale (widening of the wavelet) reveals low-frequency trends. Otherwise, high-frequency trends are obtained.

3.1.1 Continuous Wavelet Transform

The continuous wavelet transform (CWT) is determined by multiplying a scattered wavelet function by the scaling amount for the entire time [30]. Wavelet coefficients determined are the results of this multiplication. Equation (7) shows the continuous wavelet function $\psi_{s,\,b}$with displacement amount *b* and scaling amount *s*. The shift must be a positive real number, the shift factor being the real number:

$$\psi_{s,b} = \frac{1}{\sqrt{s}}\,\psi\left(\frac{t-b}{s}\right) \tag{7}$$

Equation (8) shows the CWT determined by using the wavelet function:

$$\mathrm{CWT}\,(s,b) = \frac{1}{\sqrt{s}}\int_{-\infty}^{\infty} f(t)\psi\left(\frac{t-b}{s}\right)dt \tag{8}$$

where $f(t)$ is the signal, the decomposition filters are the shift (time) and scaling parameters, and these parameters provide analyzing the signal at different frequencies. If the time parameter (b) is greater than 0, it shifts to the right. Also, the time parameter goes to the left if it is less than zero. The scaling parameter (s) increases in the time axis when it is greater than 1, and this condition causes decreasing the amplitude of the processed signal. If the function decreases, then the signal amplitude decreases in the time axis. Accuracy increases as the similarity ratio between signal and wavelet increases (correlation), while selecting mother wavelet changes for different applications.

When applying the CWT to a signal, the main wavelet is shifted along the time axis, depending on the offset value. For each shift, the coefficient of similarity is determined between the signal and the wavelet. After the coefficient finding process is continued until the entire signal is covered, the wavelet scale is changed, and the same steps are repeated for each scale value.

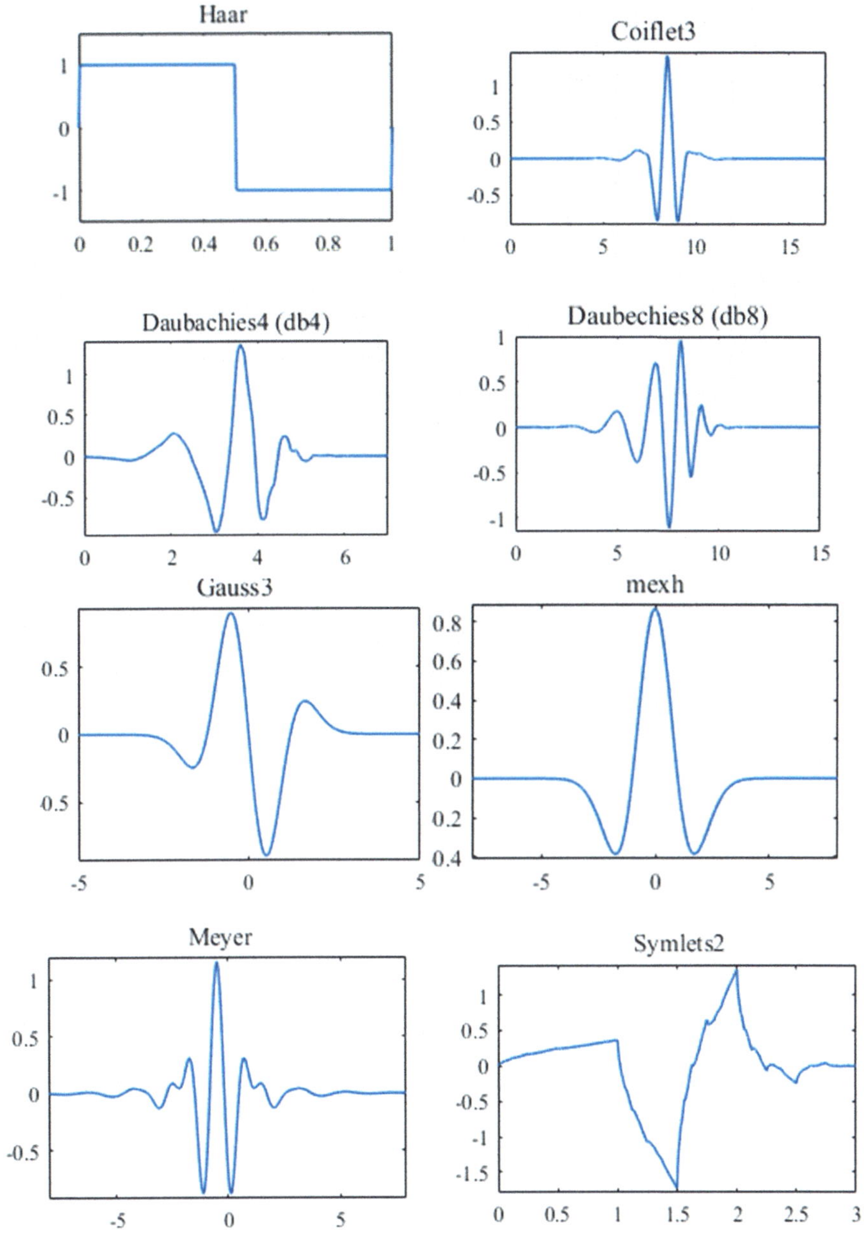

Fig. 6 Most common wavelet functions used in the applications

The transient regime event is discussed to give an example of fault detection by the CWT method. For transient regime event detection, the CWT of the signal is first taken using Haar wavelets. Then the squares of the CWT coefficients

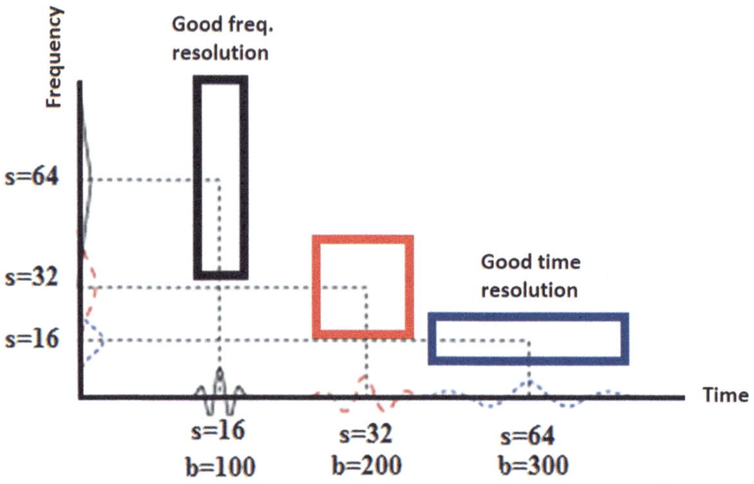

Fig. 7 Wavelet time-frequency representation

are calculated, and a scalogram (wavelet power spectrum) graph is plotted. For the exact determination of the fault point, the cumulative graph is obtained by accumulating CWT coefficients on the scale axis. In the last stage, the peak values in the cumulative plot of the CWT coefficients are determined. The determined peak values give the fault points.

In the analysis performed by the CWT method on the signal, the coefficients are greater at the time of failure and higher noise values. The coefficients have positive and negative values on all scales at fault points. In the case of noise, the resulting coefficients are positive on some scales and negative on other scales. If the CWT coefficients are accumulated in all scales, as in the cumulation graph, the noise coefficients of different scales dampen each other. At the same time, the coefficient values corresponding to the fault points are expanded. The peak values in the CWT coefficients plots correspond only to the moment of failure due to the damping of the noise coefficients.

In Fig. 8, the CWT analysis was performed for the transient voltage event. In the normalized signal, the grid frequency is 50 Hz, the selected frequency of sampling is 10 kHz, and the simulation time is 0.2 seconds. The oscillating transient regimen occurred between 0.06 and 0.65 seconds, with a frequency of 900 Hz, and the amplitude of the voltage at the time of the event increased to twice the nominal voltage. There is also a high amount of white Gaussian noise on the signal. The wavelet used for the CWT is the most basic wavelet Haar wavelet.

CWT method has a high computation time, so it is not suitable for real-time applications. As shown in Eq. (10), the closed-form solution of the integral cannot be found except in a few cases. Therefore, as in the example in Fig. 8, the CWT of a signal is first acquired at a discrete time. In the light of these values, the scale

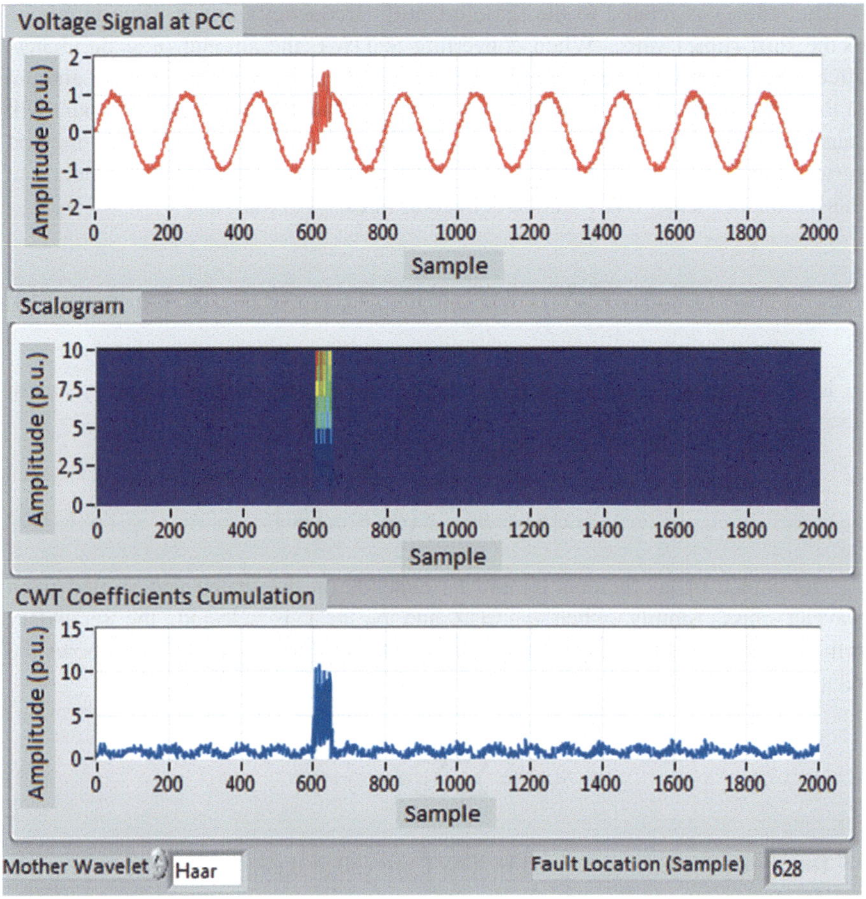

Fig. 8 Analysis of transient with CWT

and translation values for the CWT are determined, and the CWT coefficients of the sign are found.

3.1.2 Discrete Wavelet Transform

Since the CWT method requires a lot of processing load, DWT is used in applications with two and multiples of translation and scaling [31]. Equation (9) shows the DWT function obtaining from the wavelet function:

$$\mathrm{DWT}\,(s, b) = 2^{-s/2} \int f(t)\psi\left(2^{-s}t - b\right) dt \tag{9}$$

The value s is related to the scale quantity (frequency), and b value is related to the shift (time) value. When converting to DWT, the amount of scale change affects the accurate measurement in the analysis. Besides, selecting a small amount of the scale decreases the signal processing speed, and selecting a large amount will cause reducing the frequency resolution. A filter bank is used in the DWT method performing in the applications. The discrete expression for DWT corresponds to translation and scale. DWT can be defined as in Eq. (10) [32]:

$$\mathrm{DWT}\,(j,k)=\sum_{n\in Z}\sum_{k\in Z}S(n)g_{j,k}(n),\qquad g_{j,k}\in Z,\,j\in N,\,k\in Z \tag{10}$$

Here, $g_{j,\,k}(n)$ is the main wavelet, and it is also defined as limited energy or fast decomposition time function:

$$g_{j,k}(n)=a_0{}^{\frac{-j}{2}}g\left(a_0{}^{j}n-kb_0\right) \tag{11}$$

The choice of coefficients a_0 and b_0 depends on the scaled and shifted main wavelet class. Simply, when $a_0 = 2$ and $b_0 = 1$ is selected, the secondary orthonormal wavelet transform function is obtained [20]. Equation (12) shows the relationship between $S(n)$ and $g_{j,\,k}(n)$):

$$S(n)=\sum_{n\in Z}\sum_{k\in Z}d_{j,k}g_{j,k}(n) \tag{12}$$

The d_{jk} coefficients generated by DWT are called the similarity index between the main wavelet and the signal. The index has a large value, when the similarity ratio is strong, and oppositely to this condition, and the index will be small.

A series of high-pass filters are used to determine the DWT of a signal, and high-frequency components are distinguished. Similarly, a series of low-pass filters are used to distinguish low-frequency components. Information of low frequency in the signal is analyzed with approximation coefficients (cA_n), while low-frequency information is obtained with detail coefficients (cD_n) [33]. Mathematical equations of coefficients are given in Eq. (13):

$$cA\,[n]=\sum_{k}S(k)g\,[2n-k]\,;\qquad cD\,[n]=\sum_{k}S(k)h\,[2n-k] \tag{13}$$

Multi-resolution Analysis (MRA)

The high-resolution analysis method proposed by Mallat in 1988 to separate a signal into more than two sub-frequency values using a high- and low-pass filter

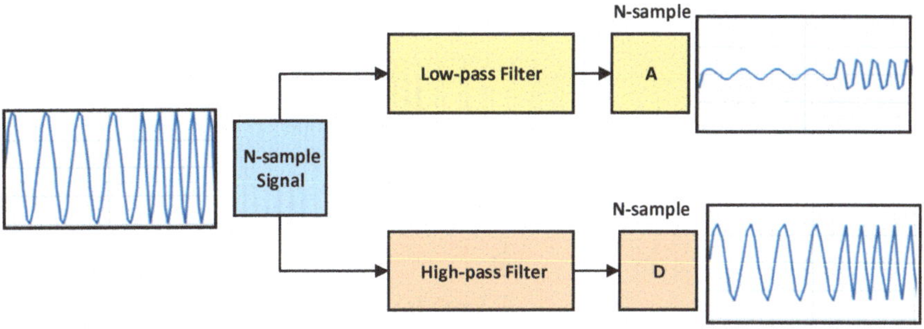

Fig. 9 Single-level multi-resolution analysis filtering

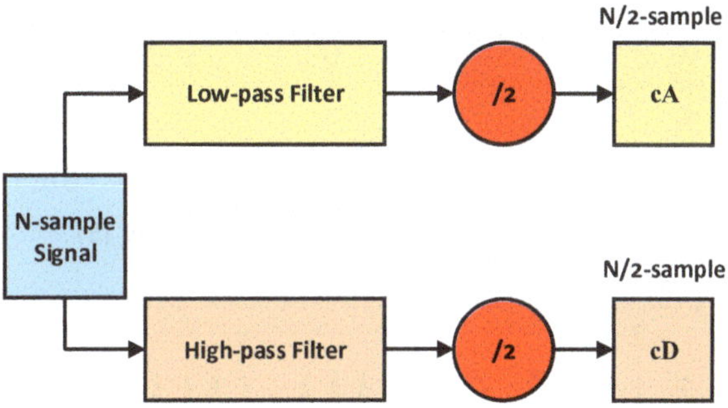

Fig. 10 Single-level multi-resolution analysis filtering and sample reduction

is still one of the most widely used DWT methods [27]. In this algorithm based on the filter bank model and performing fast discrete wavelet transform (FDWT), the approximation coefficients (cA_n) show high-scale and low-frequency signals, while detail coefficients (cD_n) are low-scale and high-frequency, as shown in Fig. 9 [30].

In MRA applications, the number of samples can be halved to avoid doubling the number of samples. The status of the approximation and detail coefficients for this process is shown in Fig. 10. Reducing the number of samples at each filter output in half can lead to the loss of information of the low-frequency components in particular in the analyzed signal.

Filter outputs in single-level MRA results for a signal with variable frequency are given in Fig. 11. The similarity ratio of the signal, low-frequency approximation coefficients at higher frequencies are presented in Fig. 11. It is also seen that the detail coefficients are higher in Fig. 11.

The desired frequency band gap is obtained by using both a low-pass filter and a high-pass filter in the MRA method. Three-level WT is performed to determine the approximation and the detail coefficients of the signal, and the sample amount at each level are given in Fig. 12.

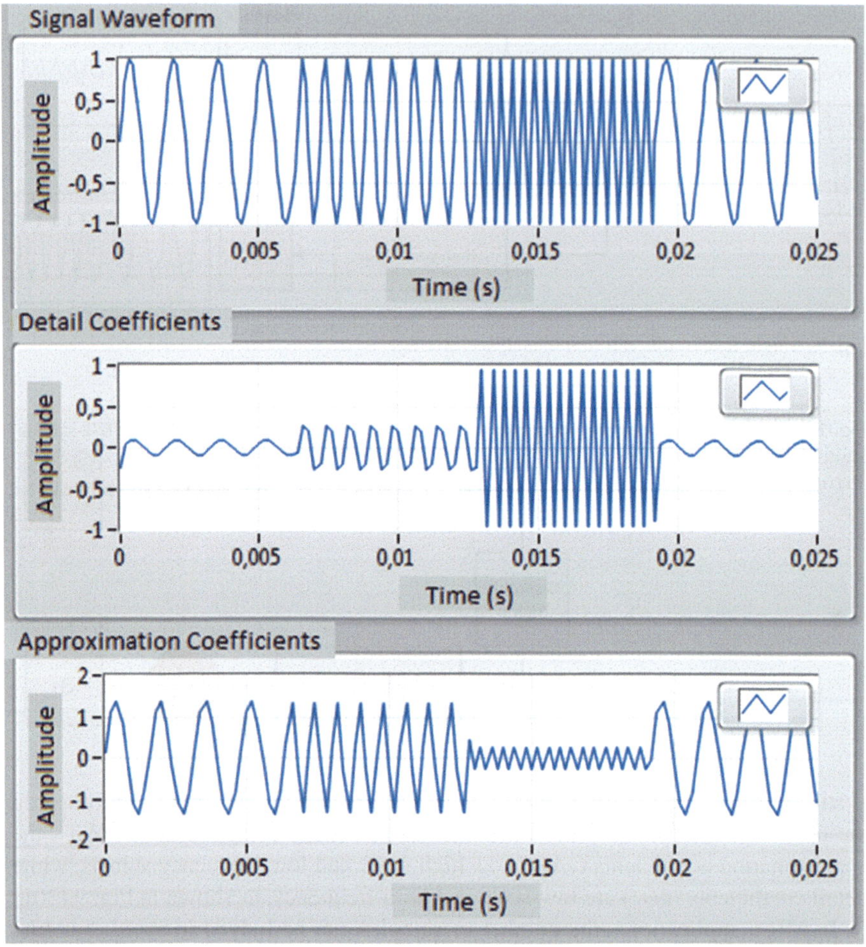

Fig. 11 Filter outputs for single-level multi-resolution analysis

3.1.3 Wavelet Packet Transform (WPT)

R. Coifman and V. Wickerhauser developed the WPT, a continuation of multi-resolution separation. As with the DWT method, filter banks are used in the WPT. Unlike DWT in the WPT method, the separation process is applied to both the detail and approximation coefficients at each level simultaneously to provide better frequency resolution. In DWT operation, the parsing process can never be applied to detail coefficients again. The WPT method, considered a hybrid method between DWT- and CWT-based techniques, provides a more efficient alternative than other methods in terms of calculation with sufficient frequency resolution.

Mathematically, the signal function $S^n(x)$ is subjected to WPT as in Eq. (14). Here $n = 0, 1, 2 \ldots, (2^{-i} - 1)$ represents the frequency parameter (node number),

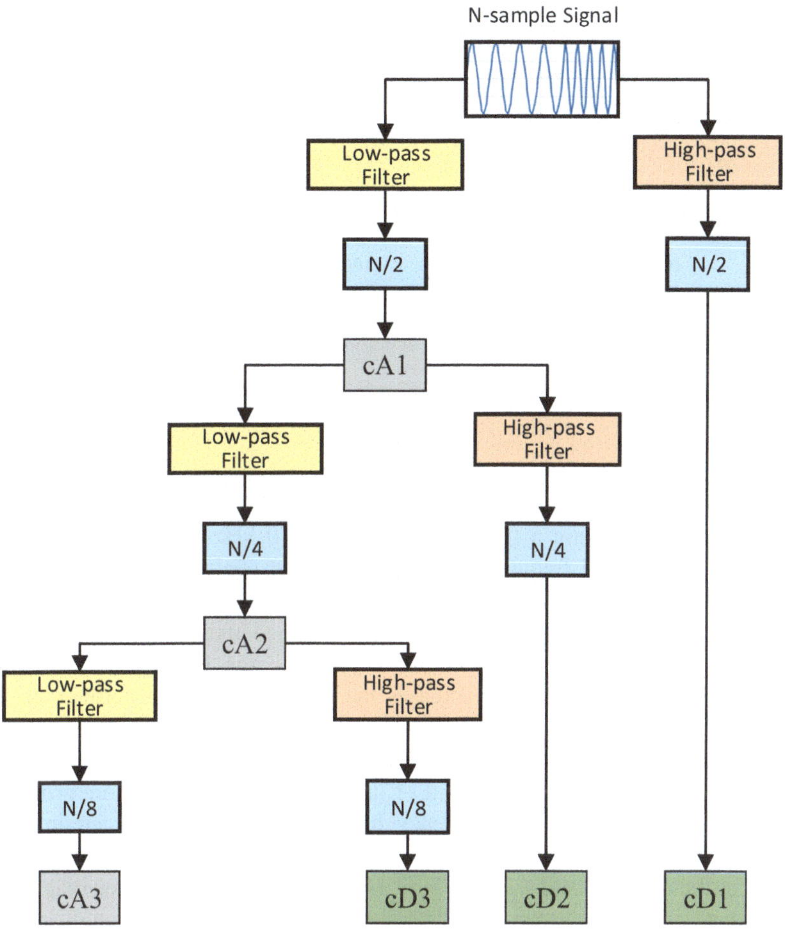

Fig. 12 Obtaining coefficients by multi-resolution analysis

and i represents the decomposition level (parameter depth) of the wavelet packet tree. Besides, the position parameter (sampling time) is indicated by j:

$$S_{i,j}^{n}(x) = 2^{-j/2} S^{n}\left(2^{i}x - j\right) \tag{14}$$

In Fig. 13, the decomposition tree diagram formed as a result of the analysis made with three-level WPT is given. The numbers in the box represent the wavelet coefficients, and the numbers on the path correspond to the process applied to the signal. 0 indicates that the signal is filtered through a low-pass filter; the number 1 indicates that the signal is filtered through a high-pass filter. As can be seen in the

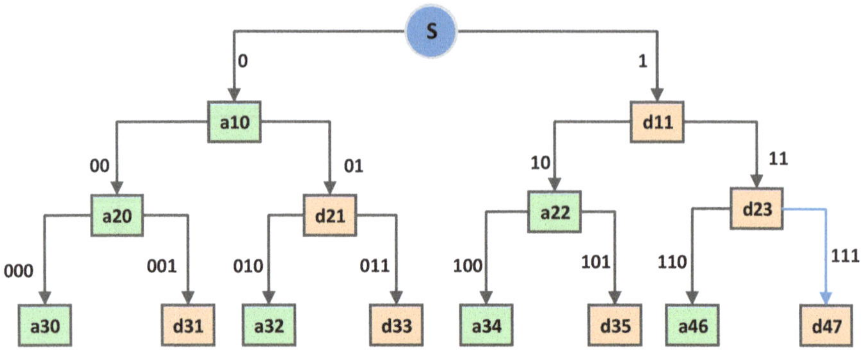

Fig. 13 Three-level decomposition tree diagram

figure, for each approximation and detail coefficients at each level, 2^n coefficients are obtained by decomposing.

For example, applying a five-level WPT with 10 kHz sampling frequency to a signal at 50 Hz grid frequency yields 32 wavelet coefficients. The bandwidth of each of the obtained wavelet coefficients is about 156.5 Hz. If the correct parameters are selected, the frequency ranges examined in the signal can be selected quite small. Therefore, WPT is frequently used in the literature, especially in harmonic analysis. Although it increases the processing load according to FFT-based harmonic analysis, it gives more accurate results.

In the analysis with WPT, the signal is filtered through a low- and high-pass filter for each level, respectively. To get rid of this load, which means more processing load and loss of time, if the frequency band to be examined is certain, signal decomposition is performed for intermediate frequencies with the desired filtering process at the desired levels only. Figure 14 provides an example tree diagram for intermediate frequency separation. The signal was first passed through low-pass filtering and then subjected to high- and low-pass filtering to obtain the coefficient a32, respectively. Intermediate frequency separation provides great convenience, especially if the frequency bandwidth is known to detect the event in fault and error analysis.

3.1.4 Undecimated Wavelet Transform (UWT)

UWT method is a popular WT-based signal processing method, and it is used generally in many image processing applications. UWT method is used in the detection of power quality disturbances in the last few years in the literature [36, 37]. The main signal is the same as the length of the coefficients due to the number of detail, and the approximation coefficient sample is not halved. The analysis carried out by the UWT method is independent of the time shift. Thus, analyzing signals

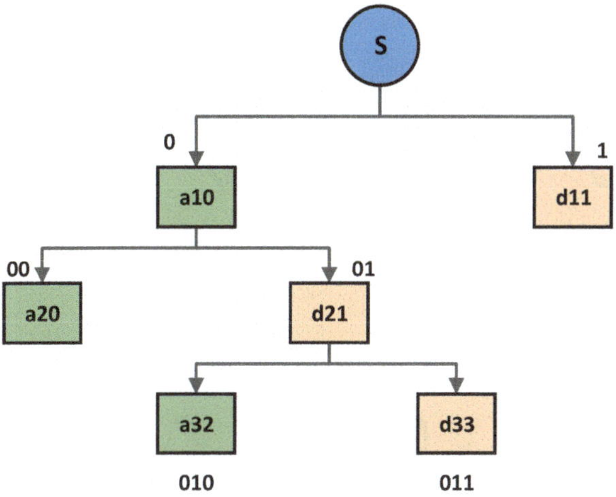

Fig. 14 Intermediate frequency signal decomposition tree diagram

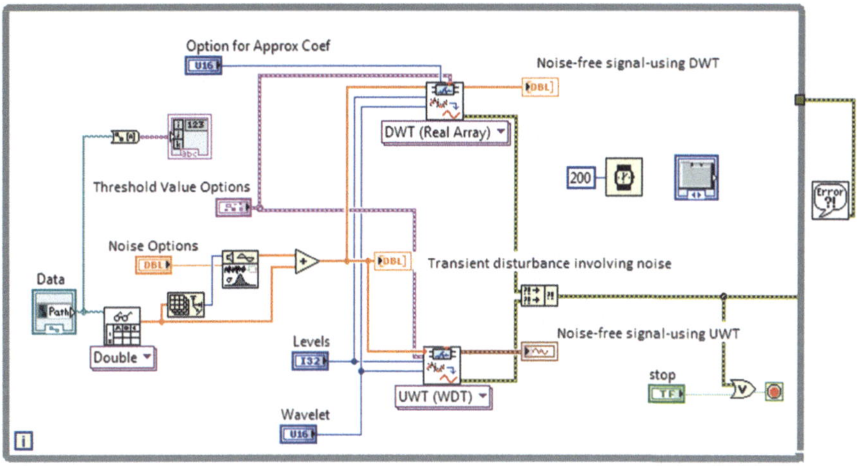

Fig. 15 LabVIEW program developed for WT-based noise reduction

in real time, reducing the noise, and detecting valley are the main advantages of the UWT method compared to other WT methods [34, 35].

The LabVIEW program in Fig. 15 has been developed to compare the UWT and DWT methods for the signal-free process. The noise signal was added to the pre-recorded transient event signal and was input to the Wavelet Denoise Express VI block for both DWT and UWT analysis. For both WT methods, the db4 wavelet was selected as wavelet and applied as six levels.

As can be seen in Fig. 16, as a result of the noise reduction process based on UWT, a more accurate signal was obtained than the DWT-based method. The UWT

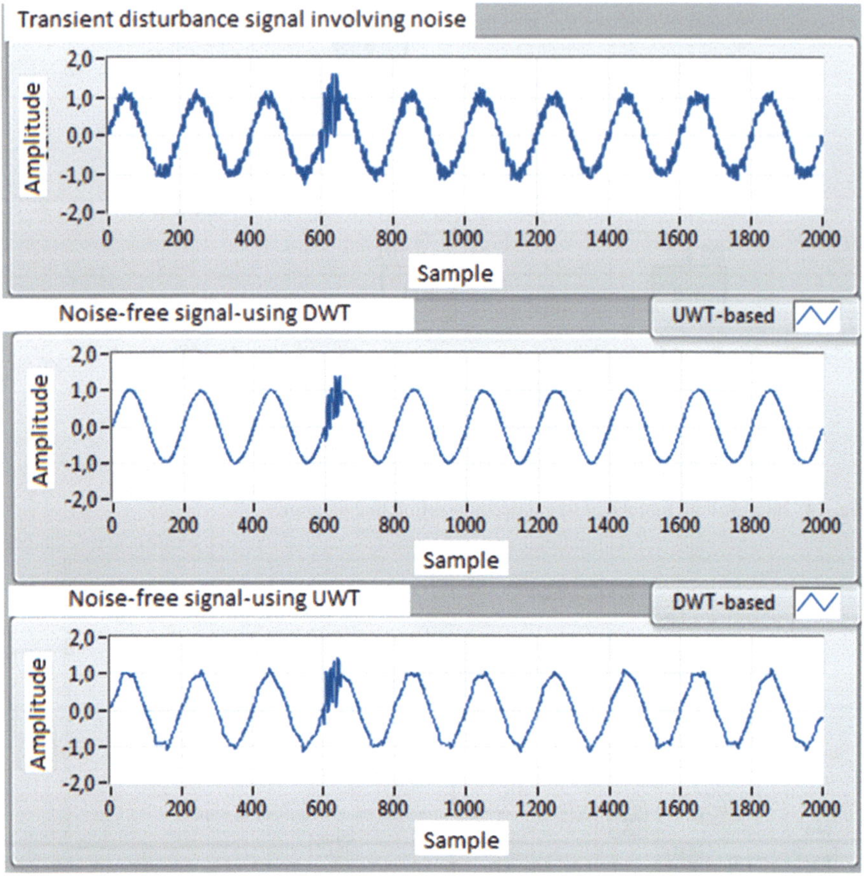

Fig. 16 Comparison of methods for WT-based noise reduction

method acts as a high performance indicated as in this condition, although it requires a little more computation processing load.

The WT method is frequently used to detect peaks and troughs on the signal. Compared to the DWT-based peak detection process, the detection process performed with the UWT gives more reliable results, and the UWT method is independent of noise. The reason for this is that the UWT-based method first finds zero-crossing points in low-resolution signals and then applies the same process to higher-resolution coefficients. Detection of zero transitions reduces noise when applied to low-resolution coefficients and increases sensitivity to detect locations with the highest amplitude when applied to high-resolution coefficients. In Fig.17, the results of UWT-based and curve fitting-based peak detection methods are compared with the signal that is observed in the transient event. The threshold value is 1.1 p.u. in this analysis, and amplitude value overlapping results were obtained by using UWT-based peak detection method.

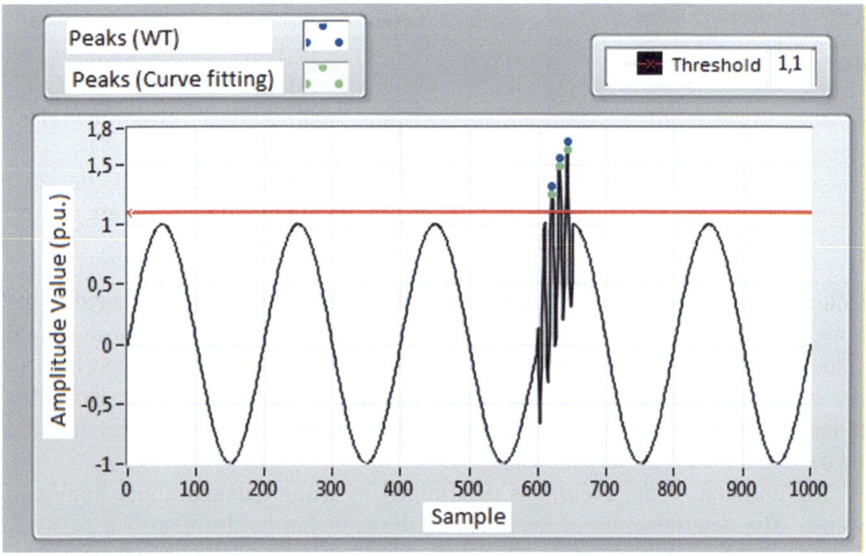

Fig. 17 Peak detection and curve fitting-based method observed in the transient event

UWT is generally used to detect power quality events in real time. Time shift is not required in UWT method and is more successful in noise purification and peak and pit detection than other methods. It is also superior to the DWT method for examining the different frequency components of the signal under examination since the number of samples is not halved after each filtering process, as in the DWT method.

4 Wavelet Transform-Based Power Quality Disturbance Detection

In this section, firstly, the structure of the WT-based event recognition system developed in real time for the power quality disturbance detection is presented. The proposed system is tested with different power quality disturbances by using WT method in real time. The real-time performance of the developed fault detection system is also evaluated with regard the fault detection time.

UWT method using five levels of decomposition by db8 is performed in the experimental study, and a data acquisition card (DAQ) acquired a voltage signal at a 10 kHz sampling frequency. Table 1 shows the obtained frequency levels of approximation and detail coefficients after the experimental analysis. The frequency range in which events such as voltage swell, voltage sag, and voltage interruption are evident corresponds to the d5 approximation coefficient; the frequency range in which transient events are apparent corresponds to the coefficient d3. Threshold

Table 1 Frequency ranges of coefficient components for five-level UWT analysis

Level	Coefficient value	Frequency range
1	Detail_1 (d1)	2.5–5 kHz
2	Detail1_2 (d2)	1.25–2.5 kHz
3	Detail1_3 (d3)	625 Hz–1.25 kHz
4	Detail1_4(d4)	312–625 Hz
5	Detail1_5 (d5)	156–312 Hz
5	Approx_5 (a5)	0–156 Hz

values were determined by using national and international criterions to calculate the power quality disturbances after the UWT analysis. The over- and under-threshold values of the determined coefficients are evaluated in the developed algorithm to detect and define the kind of power quality disturbances in real time. Peak and valley points exceeding the threshold values were used for the definition of event duration by detection the event start and end times.

The duration of the event was determined by using maximum and minimum values. The determination of event signal duration can be distinguished between event types like voltage ~~jump~~ swell, voltage sag, and events such as transient events. The difference of RMS value in the a5 approximation coefficient is compared to the defined threshold value to detect the type of power quality disturbance event.

4.1 Monitoring Nominal Voltage Signal

DAQ card is used to receive a voltage signal from the DG system, and power quality disturbances are checked for any faults by using developed fault detection algorithm. Figure 18 shows the coefficients obtained as a result of the nominal voltage waveform and five-level UWT analysis.

4.2 Detection of Voltage Sag Event

Voltage sags are evaluated at national and international standards for the event duration and the amplitude of the voltage signal. The voltage signal acquired at the sampling frequency of 10 kHz on the DAQ card has decreased by 48% of the nominal voltage at 464th ms, and this voltage drop has continued up to 665th ms. Figure 19 shows the five-level UWT analysis of voltage sag signal acquired from the experimental DG system in real time. The detection time and duration of the event are determined from the d5 detail coefficient, and the coefficient a5 also indicates the drop in the amplitude of the voltage.

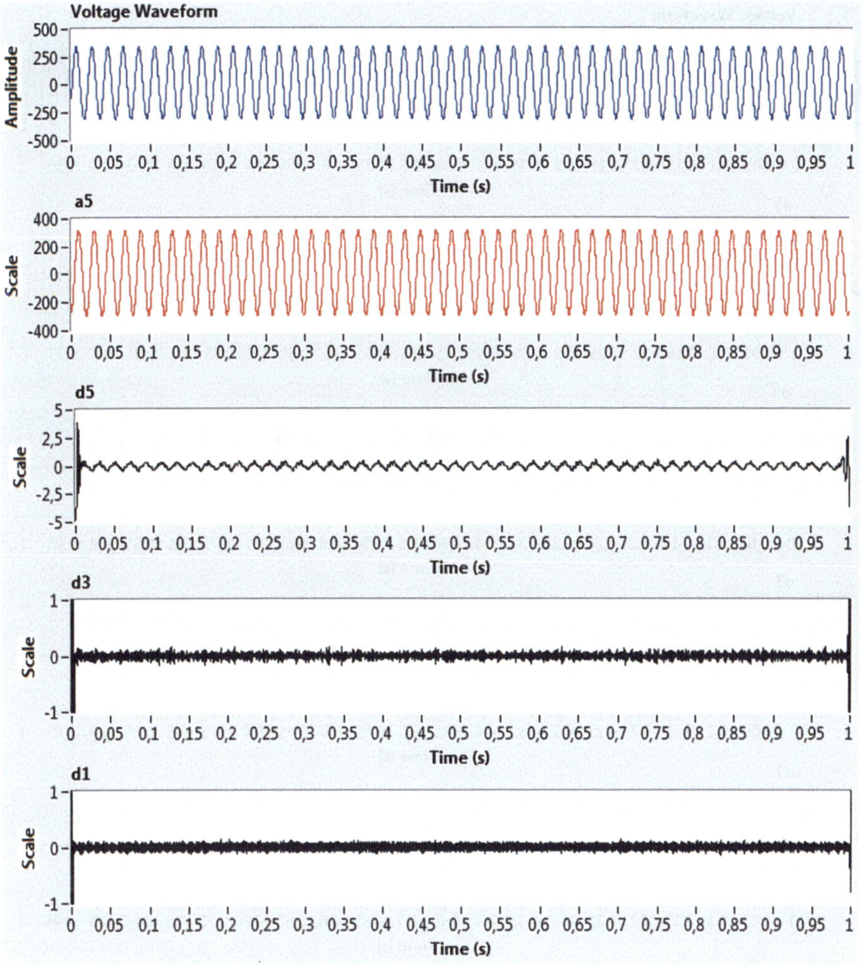

Fig. 18 Nominal voltage signal and UWT analysis

4.3 Voltage Swell Detection

The voltage swell event is an increment in the grid voltage changing between %110 and 180 of the nominal voltage, and event duration is limited between 10 ms and 60s. There was an increase in the voltage level exceeding %20 of the nominal voltage at 351st ms in the voltage swell signal acquired with a 10 kHz sampling frequency via the DAQ card, and this event lasted until 711th ms. Figure 20 shows the five-level UWT analysis of voltage swell signal acquired from the experimental DG system in real time.

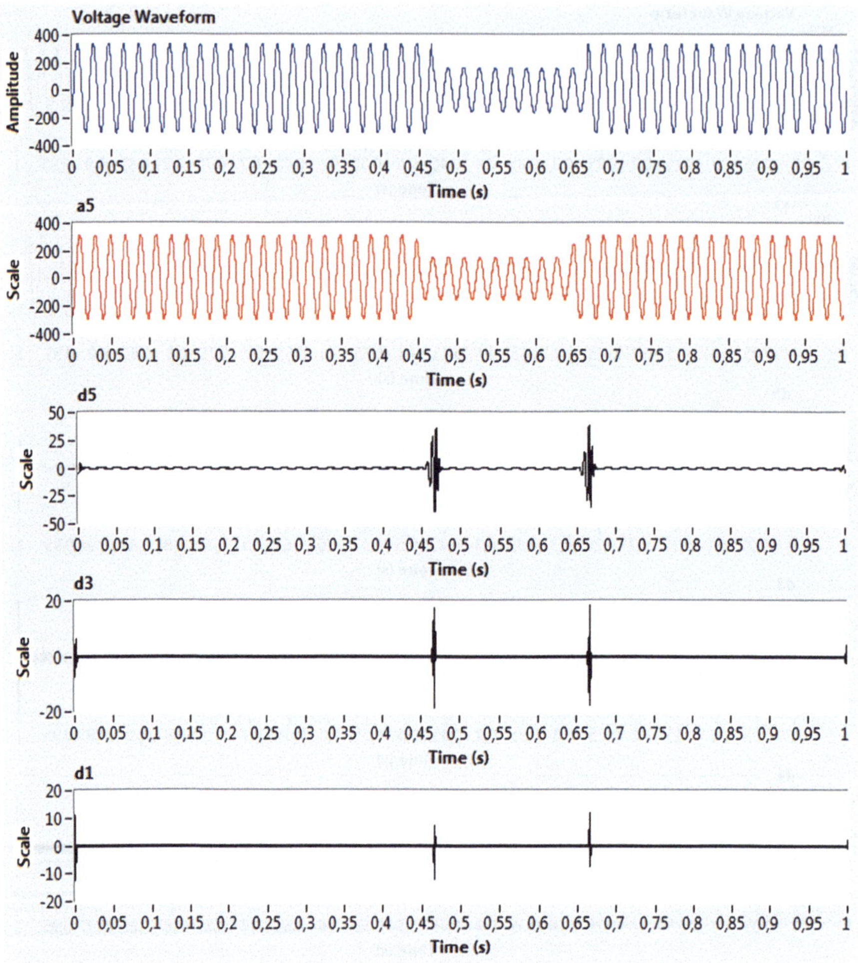

Fig. 19 Voltage sag event detection with UWT method

4.4 Voltage Interruption Detection

The voltage interruption disturbance is defined as the decrease in voltage between %90 and 100 in a DG system. The voltage interruption event is a critical fault for a DG, and the DG is in an island mode of operation in this condition. The islanded DG and all the types of equipment can be damaged due to the voltage interruption. Figure 21 shows the developed five-level UWT analysis of voltage interruption signal acquired from the experimental DG system in real time. The detection time and duration of the event are determined from the d5 detail coefficient, and the coefficient a5 also indicates the drop in the amplitude of the voltage.

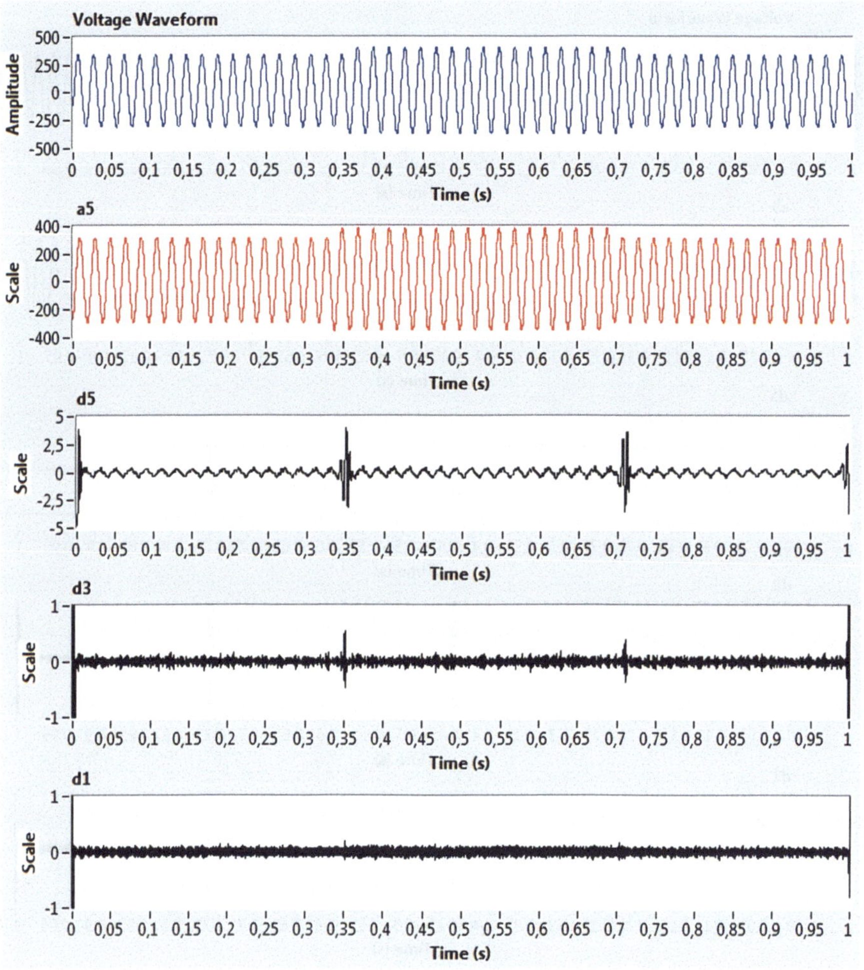

Fig. 20 Voltage swell event detection with UWT method

4.5 Transient Event Detection

In order to observe the transient event, the nominal voltage signal obtained from the signal generator is acquired in the LabVIEW environment, and then a transient event with a duration of 5 ms at a frequency of 900 Hz was added. Noise-free transient waveform and d5, d3, and d1 detail coefficient graphs obtained by the UWT analysis are given in Fig. 22. The duration of the event is detected by the d3 detail coefficient.

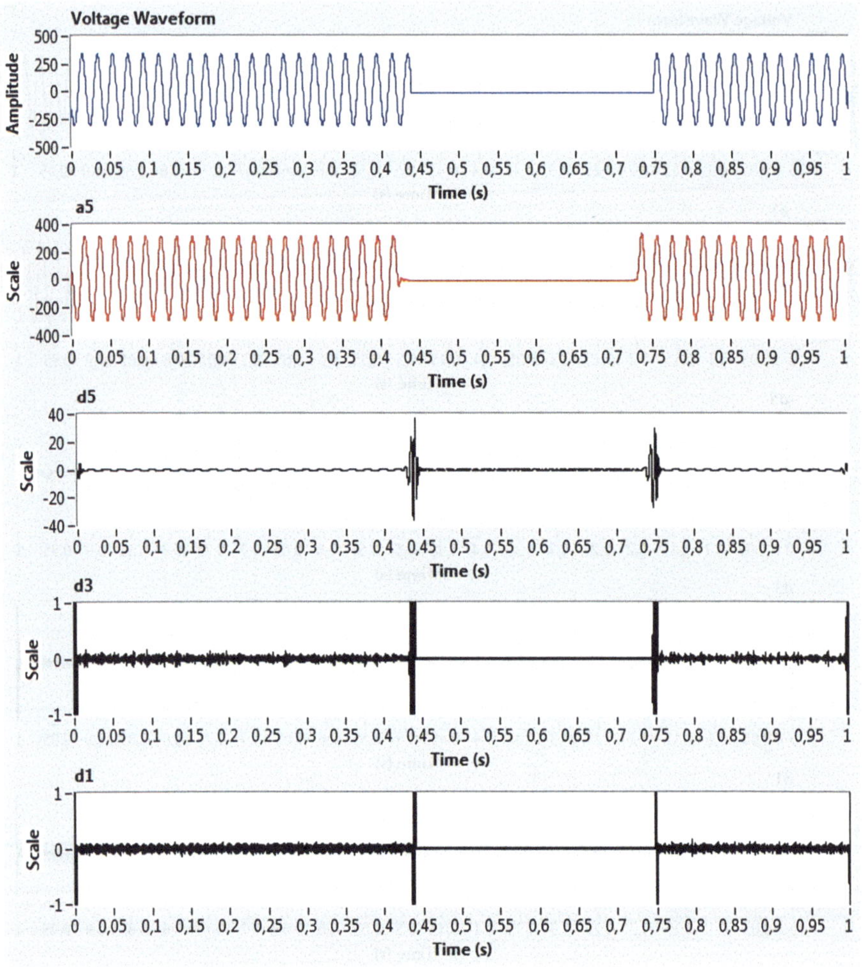

Fig. 21 Voltage interruption event detection with UWT method

5 Conclusion

Wavelet transform (WT) for the virtualization of intelligent power networks has also been experimented by researchers recently. One of the challenges for the researchers is to develop automated fault detection systems that have reliable protection units. Automated fault detection systems are required for the smart grids to provide a reliable power system.

The conventional fault detection methods are easy to implement, but they have several disadvantages, so signal processing-based intelligent fault detection methods have been proposed in the literature. Several intelligent and smart methods, which

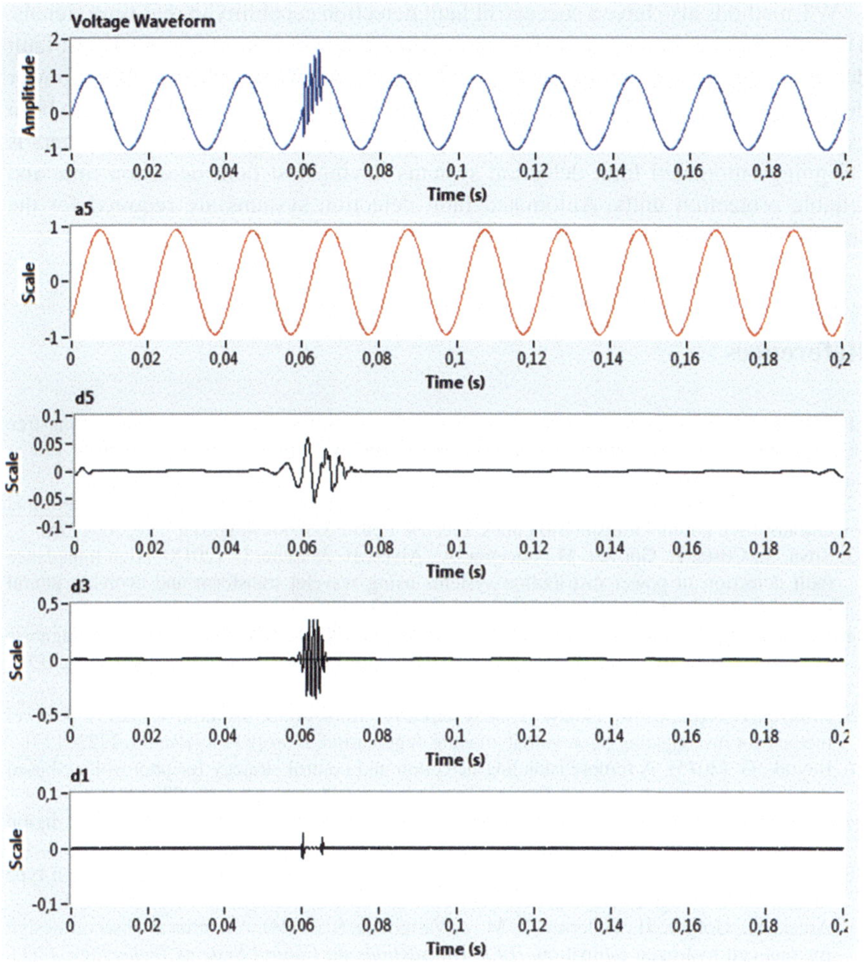

Fig. 22 Transient event detection with UWT method

include s-transform, Fourier transform (FT), and WT, have been used in the power system to detect any faults in time. Also, the WT method is useful for detecting the faults in a power system because of its high performance both in the time and the frequency domain.

WT methods are useful for detecting the faults in a power system because the WT method allows varying the window size and provides analyzing the signals in the time domain. FT only uses frequency domain analysis. Thus, WT has an advantage from FT by including time-domain analysis. Therefore, power quality disturbance signals consist of several frequency components, so the WT method presents a more reliable fault detection system comparing to other signal processing-based methods.

WT methods also have a successful fault detection capability in real-time signals. Thus, the WT method is useful for smart grid applications by using intelligent fault detection algorithms, and the researchers recently use WT method to detect power disturbances in a DG, for virtualization of intelligent power networks has also been experimented by researchers recently. One of the advantages for the researchers is designing automated fault detection systems having fast fault detection time and reliable protection units. Automated fault detection systems are required for the smart grids to provide a reliable power system.

References

1. Marques, J. P., Lazaro, C., Morais, A. P., & Cardoso, G. (2018). A reliable setting-free technique for power transformer protection based on wavelet transform. *Electric Power Systems Research, 162*, 161–168.
2. Saber, A., Emam, A., & Elghazaly, H. (2018). New fault location scheme for three-terminal untransposed parallel transmission lines. *Electric Power Systems Research, 154*, 266–275.
3. Silva, S., Costa, P., Gouvea, M., Lacerda, A., Alves, F., & Leite, D. (2018). High impedance fault detection in power distribution systems using wavelet transform and evolving neural network. *Electric Power Systems Research, 154*, 474–483.
4. Houssein, A., Héraud, N., Souleiman, I., & Pellet, G. (2010). Monitoring and fault diagnosis of photovoltaic panels. In *Energy conference and exhibition (EnergyCon)* (pp. 389–394). IEEE International.
5. Ndiaye, A., Kébé, C. M., Ndiaye, P. A., Charki, A., Kobi, A., & Sambou, V. (2013). A novel method for investigating photovoltaic module degradation. *Energy Procedia, 36*, 1222–1231.
6. Bayrak, G. (2015). A remote islanding detection and control strategy for photovoltaic-based distributed generation systems. *Energy Conversion and Management, 96*, 228–241.
7. Das, B., & Reddy, J. V. (2005). Fuzzy-logic-based fault classification scheme for digital distance protection. *IEEE Transactions on Power Delivery, 20*(2), 609–616.
8. Ferrero, A., Sangiovanni, S., & Zappitelli, E. (1995). A fuzzy-set approach to fault-type identification in digital relaying. *IEEE Transactions on Power Delivery, 10*(1), 169–175.
9. Arcak, M., Gorgun, H., Pedersen, L. M., & Varigonda, S. (2004). A nonlinear observer design for fuel cell hydrogen estimation. *IEEE Transactions on Control Systems Technology, 12*(1), 101–110.
10. Stettler, S., Toggweiler, P., & Remund, J. (2006). SPYCE satellite photovoltaic yield control and evaluation. In *Proceedings of the 21st European photovoltaic solar energy conference* (pp. 2613–2616).
11. Silva, K. M., Souza, B. A., & Brito, N. S. (2006). Fault detection and classification in transmission lines based on wavelet transform and ANN. *IEEE Transactions on Power Delivery, 21*(4), 2058–2063.
12. Zaher, A. S. A. E., McArthur, S. D. J., Infield, D. G., & Patel, Y. (2009). Online wind turbine fault detection through automated SCADA data analysis. *Wind Energy: An International Journal for Progress and Applications in Wind Power Conversion Technology, 12*(6), 574–593.
13. Burges, C. J. (1998). A tutorial on support vector machines for pattern recognition. *Data Mining and Knowledge Discovery, 2*(2), 121–167.
14. Haykin, S. S. (2009). *Neural networks, and learning machines* (Vol. 3). Upper Saddle River: Pearson.
15. Dash, P. K., Samantaray, S. R., & Panda, G. (2007). Fault classification, and section identification of an advanced series-compensated transmission line using a support vector machine. *IEEE Transactions on Power Delivery, 22*(1), 67–73.

16. Parikh, U. B., Das, B., & Maheshwari, R. (2010). Fault classification technique for series compensated transmission line using a support vector machine. *International Journal of Electrical Power & Energy Systems, 32*(6), 629–636.
17. Escobet, T., Feroldi, D., De Lira, S., Puig, V., Quevedo, J., Riera, J., & Serra, M. (2009). Model-based fault diagnosis in PEM fuel cell systems. *Journal of Power Sources, 192*(1), 216–223.
18. Kurz, T., Hakenjos, A., Krämer, J., Zedda, M., & Agert, C. (2008). An impedance-based predictive control strategy for the state-of-health of PEM fuel cell stacks. *Journal of Power Sources, 180*(2), 742–747.
19. Latran, M. B., & Teke, A. (2015). A novel wavelet transform based voltage sag/swell detection algorithm. *International Journal of Electrical Power & Energy Systems, 71*, 131–139.
20. Karegar, H. K., & Sobhani, B. (2012). Wavelet transform method for islanding detection of wind turbines. *Renewable Energy, 38*(1), 94–106.
21. Pinto, S. J., & Panda, G. (2015). Wavelet technique based islanding detection and improved repetitive current control for reliable operation of grid-connected PV systems. *International Journal of Electrical Power & Energy Systems, 67*, 39–51.
22. Naik, C. A., & Kundu, P. (2013). Power quality index based on the discrete wavelet transform. *International Journal of Electrical Power & Energy Systems, 53*, 994–1002.
23. Andrade, L. C., Oleskovicz, M., & Fernandes, R. A. (2016). Adaptive threshold based on wavelet transform applied to the segmentation of single and combined power quality disturbances. *Applied Soft Computing, 38*, 967–977.
24. Cohen, L. (1995). *Time-frequency analysis* (Vol. 778). Upper Saddle River: Prentice Hall. ISBN: 0135945321.
25. Styvaktakis, E., Bollen, M. H., & Gu, I. Y. (2000). Classification of power system events: Voltage dips. In *Harmonics and quality of power, proceedings. Ninth international conference on* (Vol. 2, pp. 745–750).
26. Jurado, F., & Saenz, J. R. (2002). Comparison between discrete STFT and wavelets for the analysis of power quality events. *Electric Power Systems Research, 62*(3), 183–190.
27. Grossmann, A., & Morlet, J. (1984). Decomposition of Hardy functions into square integrable wavelets of constant shape. *SIAM Journal on Mathematical Analysis, 15*(4), 723–736.
28. Polikar, R. (1999). The story of wavelets. In *Physics, and modern topics in mechanical and electrical engineering* (pp. 192–197).
29. Percival, D. B., & Walden, A. T. (2006). *Wavelet methods for time series analysis* (Vol. 4). New York: Cambridge university press.
30. Strang, G., & Nguyen, T. (1996). *Wavelets and filter banks*. Wellesley: SIAM. Wellesley-Cambridge Press.
31. Arı, N., Özen, Ş., & Çolak, Ö. H. (2008). *Dalgacık teorisi*. Ankara: Palme Yayıncılık.
32. Zhao, W., Song, Y. H., & Chen, W. R. (2001). Improved GPS travelling wave fault locator for power cables by using wavelet analysis. *International Journal of Electrical Power & Energy Systems, 23*(5), 403–411.
33. Zhao, W., Song, Y. H., & Min, Y. (2000). Wavelet analysis based scheme for fault detection and classification in underground power cable systems. *Electric Power Systems Research, 53*(1), 23–30.
34. National Instruments, LabVIEW 2014 advanced signal processing toolkit manual (2014).
35. Abdel-Galil, T. K., Kamel, M., Youssef, A. M., El-Saadany, E. F., & Salama, M. M. A. (2004). Power quality disturbance classification using the inductive inference approach. *IEEE Transactions on Power Delivery, 19*(4), 1812–1818.
36. Bayrak, G. (2018). Wavelet transform-based fault detection method for hydrogen energy-based distributed generators. *International Journal of Hydrogen Energy, 43*(44), 20293–20308.
37. Yılmaz, A., & Bayrak, G. (2019). A real-time UWT-based intelligent fault detection method for PV-based microgrids. *Electric Power Systems Research, 177*, 105984.
38. Bayrak, G., & YILMAZ, A. (2019). Assessment of power quality disturbances for grid integration of PV power plants. *Sakarya University Journal of Science, 23*(1), 1–1.

Monitoring and State Estimation of Modern Power Systems

Hatim Ghadban Abood

1 Introduction

Power systems are one of the most critical infrastructures so that their security and reliability play a key role in national developments. Even though computer-based systems are utilised in the energy control centres to minimise the risks of systems failure, cascade blackouts have still occurred, like that of the North-East American power network in 2003 [1]. The development of telecommunications such supervisory control and data acquisition (SCADA) systems and computer-based power applications contributed to the establishment of energy management systems (EMS) at the control centres of power systems [2, 3]. The essential power applications of EMS include the power system state estimation that was proposed by Fred Schweppes [4].

Conventional power systems contain three main stages: generation, transmission and distribution systems. These management systems are equipped with numerous measuring devices and sensors that are installed widely in the generation, transmission and distribution systems. The EMS processes the received measurements using an iterative operation of SE to find the voltage magnitudes and phase angles for each substation in a power system [4, 5]. These voltages are, subsequently, employed to determine the operating conditions of the power system. Due to the importance of this security decision, the quality of the SE solution should be maintained within acceptable limits [5]. The coordination among the different applications in the EMS replaced the function of the traditional power system operator that had the responsibility to ensure economic operation of the system and to assess the

H. G. Abood (✉)
Department of Power & Electrical Machines, College of Engineering, University of Diyala, Baqubah, Iraq
mailto: hatim.abood@engineering.uodiyala.edu.iq

M. M. Banat, S. Paiva (eds.), *Smart Technologies for Smart Cities*,
EAI/Springer Innovations in Communication and Computing,
https://doi.org/10.1007/978-3-030-39986-3_5

security in the case of equipment failure and possible occurrence of transmission line outages [6]. Figure 1 summarises the control actions for different states of the power systems, which, in turn, are based on the results of the state estimation. The flowchart shown in Fig. 1 starts with receiving the required measurements and ends with maintaining a normal and secure state using the available power applications in the EMS such as the contingency analysis and the optimal power flow (OPF) [3]. There are two main paths in the procedure of security assessment, the mainstream (green-coloured) that is directed to the ultimate goal of a normal and secure state and the lateral branches (yellow-coloured) that are related to the abnormal and insecure states. The main operating security states of a power system and the relevant control decisions are:

- Normal and secure
- Normal, but insecure
- Emergency state
- Restorative state

Thus, the power system state estimation is the fundamental operation of monitoring and control activities in energy control centres.

Based on the estimated states, subsequent actions are taken at the control centre. Power system state estimation is an essential part of the core module of the security analysis system in energy control centres. Hence, power system state estimation has been incorporated in the packages of the EMS. Even though the main function of the state estimator is to provide the best guess of the state of the power system, it is functionally involved in most of the power applications in the energy control centres. Figure 1 demonstrates the coordination among the different applications of the EMS at the control centres. Thus, the EMS has efficiently replaced the functions of the traditional power system operator of ensuring economic operation and assessing the security of the system in the case of equipment failure and transmission line outages, which is shown as a subsystem of the data acquisition and state estimation (the lower block in Fig. 1).

Figure 2 shows that the state estimation is the key operator of other monitoring and control activities (not only the decision of normal/emergency operation). The interconnected power applications shown in Fig. 2 should benefit from recent technologies in measuring devices and sensors (e.g. PMUs) as well as the mathematical/statistical criteria of the horizontal and vertical scopes of the problem. However, Fig. 2 includes the conventional power system state estimator that is based on the SCADA system.

Based on the interconnected functions of Fig. 1, the rest of the chapter discusses the characteristics of measurement set that is required for an accurate and reliable state estimation. The chapter also addresses the challenges that affect the numerical performance of the state estimation and discusses the available and the proposed solution methodologies.

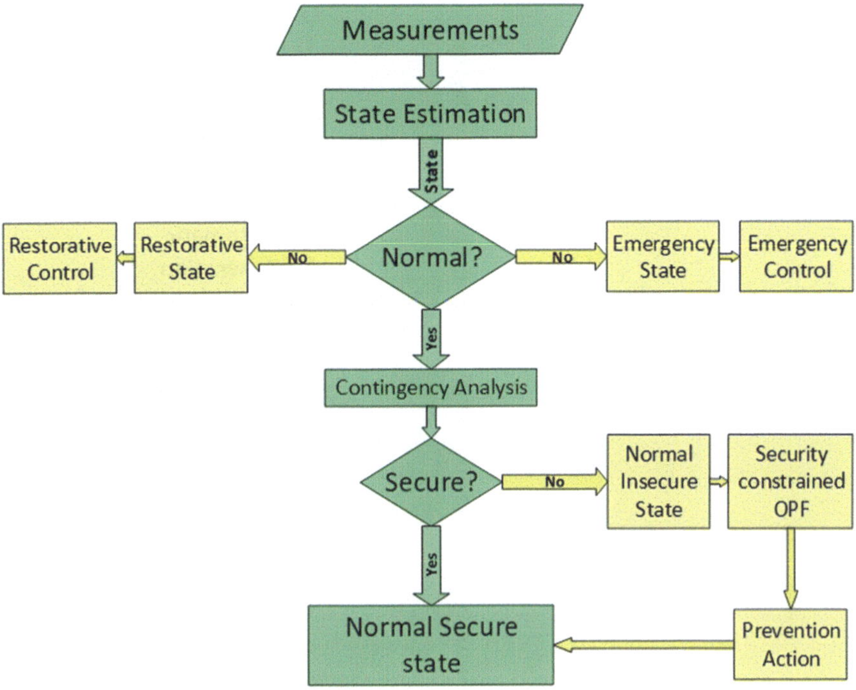

Fig. 1 Functional diagram of the static security assessment

2 Why State Estimation?

It is essential to estimate the existing state of power systems before applying any control action. However, before the advent of the state estimation, the power system operator had responsibility for most of the real-time control centre functions. The actual events faced by the operator were occasionally unexpected and had not been included in the planning cases. The operator used to depend on a large number of load-flow results. However, in practice, the telemetry measurements cannot satisfy the requirements of conventional load-flow studies mainly due to the following reasons [2, 4]:

- The inputs to the conventional load-flow programs are confined to the injection powers (P_{inj} and Q_{inj}) at load buses and P_{inj} and the voltage magnitude $|V|$ at voltage-controlled buses. If only one of these measurements is unavailable, the load-flow solution cannot provide the required solution, which means the state of the system will be missing. In practice, the availability of the measurements is not restricted to the requirements of the load-flow studies.
- The normal situation in power systems is that there are more measured quantities than the unknown state variables. The deployment of the available meters in the

Fig. 2 Diagram of the EMS functions

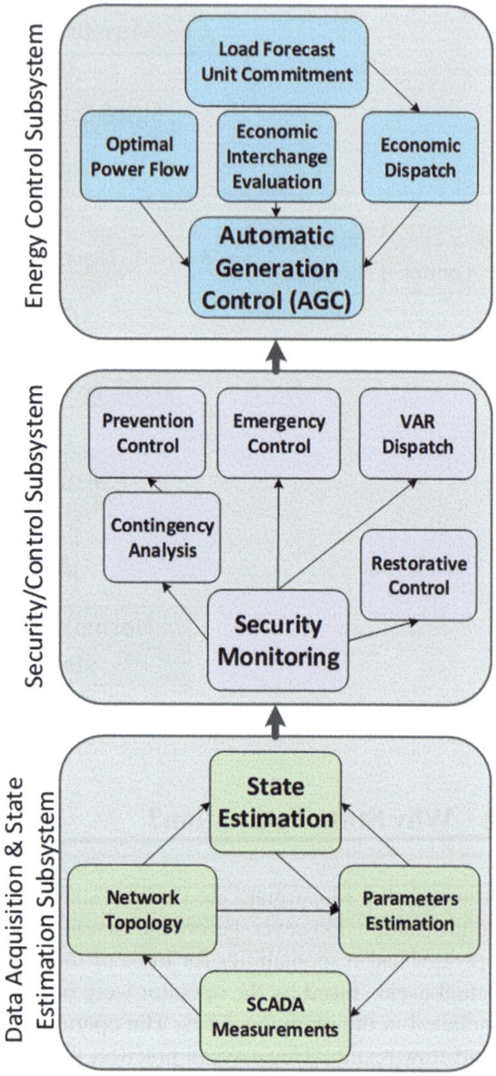

power system is not restricted to the requirements of the load-flow studies as there could be many meters at few buses in the system, while on the other hand, no—or only a few—meters in other buses. The additional measured quantities include the real and reactive power flows in the transmission lines of the power system. In the context of the load-flow applications, the ‘extra’ measurements are ‘useless’. In contrast, in the context of the state estimation, the extra measurements which are called ‘redundant’ measurements involved functionally in enhancing the state estimation observability and reliability.

- The imperfections of the available measurements and the gross error can turn the load-flow results to be ineffective.

On the other hand, it is neither economic nor technically feasible to install real-time measuring devices at every bus of the system to measure all the electric quantities required for the activities of power systems' operation and control [7, 8].

The above limitations motivated researchers to develop an alternative power application that is more efficient and less selective in dealing with the available measurements of the real power systems. Therefore, the power system state estimation has been developed to use the available set of measurements to estimate the power system state as accurately as possible [9, 10]. On the other hand, the function of the state estimation is not only that of introducing the state estimation solution; rather it includes [4, 5]:

1. Topology processing: It gathers the status data from the circuit breakers and switches placed in the whole power system to configure the diagram of the system network.
2. Observability analysis: It ensures whether the unique state estimation solution can be achieved or not, with the available set of measurements. If the system is not observable, it identifies the unobservable branches and the observable islands in the system. The conventional measurement set includes real-time measurements, virtual measurements and pseudo-measurements [5].
3. State estimation solution: It provides an estimate for the states of the power system by processing the available set of measurements and the network model. The estimated state vector includes the complex voltages of all the system's buses.
4. Bad data processing: It detects the bad data in the measurements, identifies the bad data and processes or eliminates the bad data [10–15].
5. Parameter estimation: It identifies the most likely parameters in the network. These functions are integrated into the state estimation software of the EMS at the control centre. An accurate real-time network model is crucial for proper functioning of the energy market and for the operators to detect insecure states related to the system and to assign the corrective control action appropriately.

The above functions imply many challenges and research tasks. Figure 3 describes, in brief, the main stages of the state estimation and indicates the main challenges that are associated with the power system state estimation of modern power systems. Measurement-based and statistical problems and, thus, utility decisions and operator actions should be carried out to obtain a reliable state estimator.

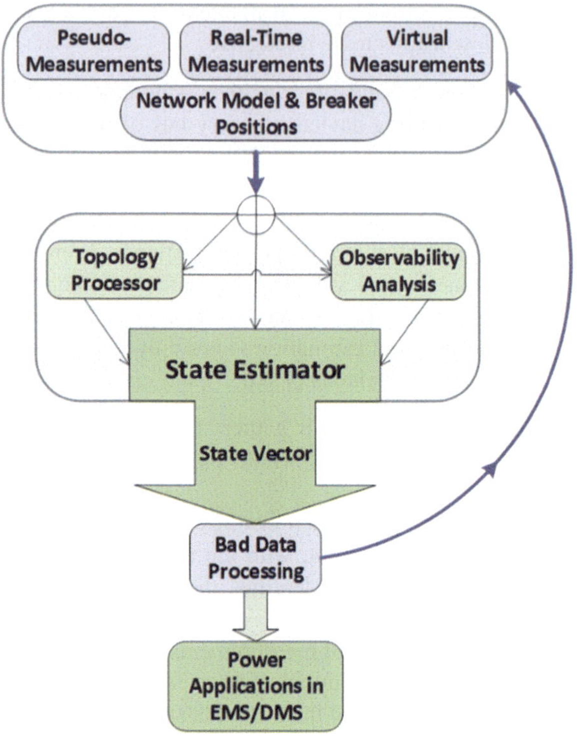

Fig. 3 The main factions and process of the power system state estimation

3 Real-Time Measurements

The real-time measurements that are known as conventional measurements are provided by meters rather than the PMUs. The conventional measurements include the active/reactive power injection measurements, active/reactive power flow measurements, voltage magnitude measurements and current magnitude measurements, if available, in addition to the tap position of the transformer which estimates the network topology. Thus, the entire list of the measurements can be utilised for power system state estimation which consists of conventional real-time measurements, synchronised phasor measurements, zero-injection measurements and pseudo-measurements.

The modern state estimation relies on the PMUs due to their ability to measure the voltage magnitudes, the phase angles and the current flow of all the incident branches with a relatively high speed (around 1 microsecond) and sufficient accuracy [16]. The significant feature of the high precision time references is provided by the Global Positioning System (GPS). The synchronised measurements of the different substations set to a common reference with a common clock (the global clock) [16, 17]. Instead of the RTUs, the synchronised phasor measurements are gathered at the substation using the phasor data concentrator (PDC). However, the relatively high cost of PMUs limits their deployment. Therefore, the dominant situation is to use hybrid schemes containing both RTUs and PMUs for efficient meter configuration [17, 18].

4 Formulation of the State Estimator

The operation of the power system state estimation contains pre and post-processing steps. The pre-estimation step is to prepare the telemetry and non-telemetry measurements and the real model of the power network. The post-estimation process includes the assessment of the state vector accuracy, bad data analysis and rechecking the observability of the system.

In the context of nonlinear regression, if there are m observations that form the $(m \times 1)$ observation/measurement vector, y that contains noises or an error vector e, the state estimation implemented to deduce the statistical relationship between the nonlinear set of measurements and the state variables needed is [19]:

$$y_i = f_i(x) + e, \quad i = 1, 2, \ldots, m, (m > n)$$

where x is the estimated response vector and $f(x)$ is a nonlinear function that is related to the observations with the unknown state variables. However, the final regression solution using the normal equations (NE) is given as:

$$\hat{x} = \left(F^T F\right) F^T y$$

where $\hat{x}$ is the estimated response (the unknowns or the state variables).

Likewise, when the following vectors are available for a power system of N buses [4, 5]:

$$z \text{ is } (m \times 1) \text{ measurement vector}$$

$$x \text{ is } (n \times 1) \text{ unknown vector}$$

$$e \text{ is } (m \times 1) \text{ error vector}$$

where $n = 2N - 1$ the state vector that includes the voltage magnitudes $|V|$ (expect that of the reference bus) and all the phase angles θ. Then, the state variables can be estimated by the following measurement equation [3]:

$$z = \begin{bmatrix} z_1 \\ z_2 \\ \vdots \\ z_m \end{bmatrix} = \begin{bmatrix} h_1(x_1, x_2, \ldots, x_n) \\ h_2(x_1, x_2, \ldots, x_n) \\ \vdots \\ h_m(x_1, x_2, \ldots, x_n) \end{bmatrix} + \begin{bmatrix} e_1 \\ e_2 \\ \vdots \\ e_m \end{bmatrix} = h(x) + e$$

where $h_i(x)$ is a nonlinear function relating the state vector x to the ith measurements.

The above equation can be rewritten for the ith measurement as follows:

$$z_i = h_i(x) + e_i \tag{1}$$

where the errors are assumed to be uncorrelated and normally distributed. Hence, (1) can be rewritten to express the measurement residuals as follows:

$$r_i = z_i - h_i(x) \tag{2}$$

Then, the objective function is based on minimising the difference between the estimated state vector and the measured values, i.e. obtaining minimum residuals. For this purpose, the weighted sum of squares of the residuals should be implemented for the whole set of m measurement [3]. Thus, the state estimate $\hat{x}$ is defined to be the value of x which minimises:

$$J(x) = [z_i - h_i(x)]^T R^{-1} [z_i - h_i(x)] \tag{3}$$

where R^{-1} is the reciprocal of the measurement variances (σ^2) which are also represented by the weights assigned to each measurement. High weighting factors are associated with the accurate measurements of small variance, and the low weights are assigned for the measurements of low accuracy. Hence, the values of the diagonal elements are given as $R = \text{diag}\left\{\sigma_1^2, \sigma_2^2, \ldots, \sigma_m^2\right\}$.

The solution of (3) is called the weighted least squares (WLS) estimator for x, and it must satisfy the following first-order optimality function [9]:

$$\partial J(x)/\partial x = 0 \Rightarrow H^T(x)\, R^{-1}\, [z - h(x)] = 0 \tag{4}$$

where $H_{ij}(x) = \partial h_i(x)/\partial x_j$ is the $(m \times n)$ measurement Jacobian matrix and $H^T(x)$ is the transpose of the Jacobian matrix. The measurement Jacobian matrix of the state estimation is different from that of the Newton-Raphson load-flow method. The Jacobian matrix is composite of submatrices corresponding to the different measurement types incorporated in the measurement set [3].

The goal is to obtain the state vector $\hat{x}$ that minimises $J(x)$ and satisfies (4). The most effective approach to do so is to implement an iterative Newton-Raphson (NR) process. Since $h(x)$ is a nonlinear function, the convergence of the NR neglects the terms of second-order derivations of $h(x)$ using the Taylor series expansion. Thus, the linear system comprising n equations must be solved iteratively via the normal equations (NE) approach. The NE of the WLS estimator for the kth iteration is as follows:

$$H^T(x)WH_x\,\Delta x^k = H^T(x)\,R^{-1}\left[z\text{–}h\left(x^k\right)\right] \tag{5}$$

$$G\left(x^k\right)\Delta x^k = H^T(x)\,R^{-1}\left[z\ \text{–}h\left(x^k\right)\right] \tag{6}$$

$$G\left(x^k\right)\Delta x^k = H^T(x)\,R^{-1}\,\Delta z^k \tag{7}$$

where $G(x^k) = H^T(x)R^{-1}H(x)$ is the *gain matrix* of the NE, $\Delta z^k = z\ - h(x^k)$ is the measurements residual of the kth iteration and $x^k = x^{k-1} + \Delta x^k$ is the increment of the state vector at the kth iteration. Then, the state vector is updated after solving the equation system to be $x^{k+1} = x^k + \Delta x^k$. For simplicity, the symbols H and G are used to refer to the Jacobian and gain matrix, respectively.

The gain matrix is a square, symmetrical, sparse matrix and a positive definite if the H matrix is of full rank. Additionally, for an observable system with sufficient measurement sets, the gain matrix should be positive definite and symmetrical [6, 9].

5 Algorithm of the Conventional State Estimation

Based on the above derivations, the main steps of solving the WLS state estimator by the NE approach are as follows [2, 10]:

1. Prepare the measurement set and the corresponding weighting factors.
2. Initialise state vector $x = x^0$ with the flat start profile where all voltage magnitudes are equal to unity and the phase angles are ($\theta = 0°$). Set iteration counter to $k = 0$.
3. Construct the Jacobian matrix, H.
4. Compute the gain matrix $G(x^k)$.
5. Calculate the right-hand side of (5).
6. Decompose $G(x^k)$ using Cholesky triangular factorisation.
7. Solve (5) for Δx^k using the forward/backward substitutions.
8. Check the convergence rate; if (max $|\Delta x^k| \le$ tolerance rate), stop; otherwise, continue.
9. Update the state vector ($x^{k+1} = x^k + \Delta x^k$) and the iteration counter ($k = k + 1$) and return to step 3.

The above procedure is repeated until the minimum error is achieved or reaches the maximum iteration assigned to the state estimation solution.

6 Challenges and Motivations

The state estimation process has a hybrid nature that is based on monitoring infrastructure of the power systems, utilities and operators' efforts and planning studies. However, although the EMS has been utilised for few decades, utilities had not been enforced to employ highly accurate real-time meters and the latest data analysis applications [11].

The Venn diagram of Fig. 4 summarises the outcomes of the first stage in the hierarchical functional scheme. However, in addition to the main blocks of the diagram, the chapter also concerns the common regions that point out the mutual relationships and the interference between different numerical and technological challenges.

The concepts are shown in Fig. 4 ranging from the base—where utility efforts are required to manage the measurements and the correct network model to the peak—where the operator's efforts at the control centres are crucial along with the programmers and data analysts.

The second stage of this research is to assign suitable actions to the challenges after acquiring the detailed view to the horizons of the potential solutions. Proposing

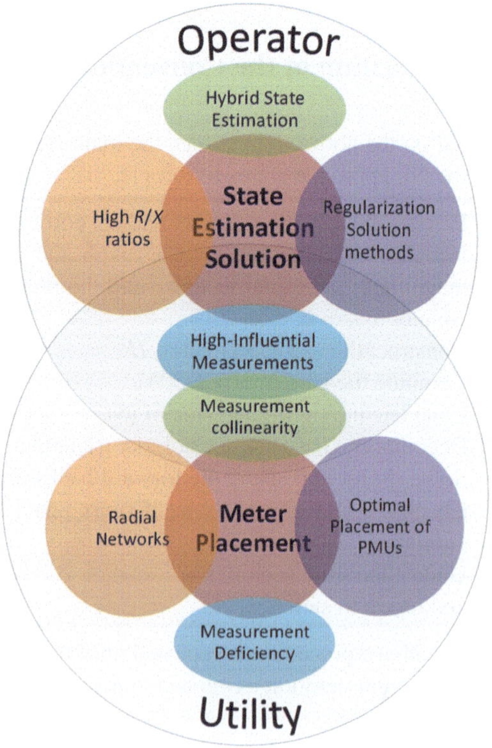

Fig. 4 Venn diagram for the main challenges of enhancing the state estimation

and implementing new methodologies and techniques to enhance the state estimation is the approach of the second stage of this research.

The key challenges for obtaining a high-performance state estimator can be classified into two categories as follows.

6.1 Challenges in Measuring Infrastructure (Utility Side)

Although the state estimation is a software package in the EMS, the data to be processed using a regression procedure for estimating the system's state is produced by physical meters that are deployed in the substations and power lines of the power systems. Hence, the adequacy, diversity and accuracy of the measuring and telecommunication systems are the primary features to be assessed for evaluating the performance of the state estimation. An inappropriate measurement set is likely to affect the convergence of the state estimation solution and deteriorate the accuracy of the estimated states.

The motivations related to measurement infrastructure have the following key challenges.

6.1.1 Measurement Adequacy and System's Observability

The minimum number of existing measurements in a power system should be sufficient to make the system completely observable and for the proper operation of the state estimator. However, this is a necessary but not a sufficient condition as a reliable state estimator requires more measurements to cover the potential problems of meter failure and line outages [12–14]. Hence, there should be a measurement redundancy. The measurement redundancy is achieved when the ratio of the available measurement to the number of the state variable is larger than unity. This redundancy is necessary because the measurements are sometimes grossly in error or unavailable owing to malfunctions in the data-gathering systems. Redundant real-time measurements are also crucial for the process of detecting and eliminating the bad data. When the observability is incomplete, the state estimator must be extended to include the unobservable pockets of the system through utilising pseudo-measurements that are based on historical data. The accuracy of the pseudo-measurements may be questionable when not updated regularly to reflect current conditions of the system.

In performing state estimation for an unobservable part of the network, it is possible to corrupt the states estimated from telemetry data. Nevertheless, the challenges are related to the low accuracy of the pseudo-measurements and the problem of rising linear dependency among the redundant measurements. The measurement redundancy may distort the state estimation solution when more than 1 meter provides the same measurements where there are repeated measurements, i.e. when measuring the same quantity more than once [20, 21]. The other challenge, which is related to the state estimation of distribution systems, is that of measurement deficiency. The distribution grids usually depend on the meters of current magnitudes (the ampere measurements), and therefore, there is a lack of power measurements in the distribution system [7, 22]. Therefore, intensive studies are still required into enhancing the performance of the state estimation in the distribution systems.

On the other hand, after employing the PMUs for the state estimation, the redundancy of the satellite communication channels became part of the adequacy calculation as those channels might be limited in many cases [23, 24]. This drove the research compass to address the availability of the phasor measurements that is subjected to a deficiency in the communication channels. Hence, additional constraints need to be considered in the fitness function of the placement algorithms.

6.1.2 Measurement Consistency

Statistically, the measurement consistency is associated with the ratio of the numbers of unknowns (the state variables to be estimated) to the number of observations (the measurements). Hence, the estimation problem can be underdetermined, exactly determined or overdetermined if the number of the measurements is less than, the same as or larger than the state variables, respectively [25, 26]. However, the predominant nature of the power system state estimation is to be overdetermined as utilities tend to install meters frequently for serving other applications such as power system protections. Nonetheless, this subsection discusses the physical consistency of the meters and the measurement sets, i.e. the infrastructure of the state estimator.

Apart from the measurement adequacy, even though the existing meters could be sufficient to achieve an observable and solvable state estimator, the available meters and their measurements differ from each other in many ways that include but are not limited to:

1. Different types of real-time inputs: the state estimator uses *data* and *status* information. The on/off status of switching devices such as the circuit breakers and transformer taps determines the network configuration. The analogue data in the form of megawatt and megavar flows and injections are monitored in the system's lines and buses every few seconds [3, 4].
2. Different measuring technologies: the measurement set of the state estimation includes conventional real-time meters, PMUs and different communication channels and historical data [7, 22]. The disparity among metering technologies includes telecommunications as the transmission of the data can be achieved using SCADA systems or using the GPS in the case of the phasor measurements [13, 24].
3. The different accuracy of the measurements: the collected measurements contain highly accurate measurements such as the phasor measurements and the virtual measurements along with pseudo-measurements that are based on untrusted

historical data [25]. Hence, different weight factors must be assigned to the individual measurements in the measurement set [25, 26].

4. Different estimation approaches: the inclusion of the phasor measurements has inspired the use of nonconventional state estimation solution methods such as linear state estimation that is based on the phasor measurements solely and the hybrid state estimator that employs both the conventional power meters and the PMUs [27–30]. Sufficient PMUs can facilitate the non-iterative linear state estimator. The PMU-based state estimator assumes that the phasor measurements are error-free and each bus in the power system is observed by one or more PMUs which is not affordable in many cases. However, the linear state estimator sacrifices the state estimation accuracy in favour of obtaining a fast state estimator. Using a direct estimator for processing nonlinear measurements may result in an approximate solution [27, 29]. Furthermore, the linear state estimation requires more measurements to implement the solution, which adds burden on the distribution system state estimation [29].
5. Uneven deployment of the meters: most of the power systems contain buses and lines containing more meters than other buses and lines in the system which yields numerical problems that affect the solvability of the state estimator [30]. Moreover, the common meter placements methods tend to install new meters in the buses that have more branches/lines [30–32]. However, this affects the distribution of the meters in the system while unnecessary for enhancing the state estimation.

6.1.3 Meters Quality and Diversity

In addition to the measurement sufficiency that is required to achieve a complete observability and solvable state estimator and the measurement consistency that is necessary for avoiding numerical stability problems, the diversity is also necessary to support the estimation of both the voltage magnitudes and the phase angles [33, 34]. A diverse set of power measurements and voltage measurements is required for estimating both voltage magnitudes and phase angles at the same level of accuracy and avoiding repetition of the measurements that may create numerical problems [35, 36]. However, the numerical problems that emerge from inappropriate measurement set range from the presence of collinear measurements [35] and high-influential [16] and bad measurements to the problem of rank deficiency of the Jacobian matrix [37]. Those data-based problems deteriorate the numerical stability of the power system state estimator and can result in ill-conditioned state estimators.

On the other hand, the distribution systems have a problem with the similarity of the available meters as most of the available measurements are ampere measurements that are produced by current meters [33, 34]. The distribution grids lack power meters that support the estimation of the phase angles. However, researchers have tried to modify the behaviour of the ampere measurements [38].

The dependence of the distribution system on the ampere measurements and the paucity of conventional power meters and PMUs lead to significant problems to the numerical stability, the convergence of the state estimator and unreliable phase angles estimates [9, 17]. Nevertheless, there is still a research gap in the configuration of the meters in the power systems. In this chapter, the terminology of meter configuration refers to the types, numbers and locations of the meters that produce the measurement set of the state estimation.

6.1.4 Bad Data Analysis

The process of bad data detection (BDD), identification and elimination is mainly associated with the operator's efforts as it is a post-estimation process based on the quality of the estimated states [11]. Nevertheless, preparing the sufficient measurement redundancy with reasonable accuracy and the strategic allocation of the meters dramatically influences the decisions of the power systems' operator. Hence, few strategic buses should have local redundancy; otherwise, incremental meters need to be utilised. However, with the advent of smart grids, the bad data processing became necessary not only for detecting bad data but also for preventing security attacks and false data injection (FDI) [12–15]. The security attack by injecting false data is one of the common challenges in the operation of smart grids [13].

This chapter aims to intensively examine the detection methods of the bad data, discover the relationships between the high-influential and collinear measurements and develop technical and analytical solutions to facilitate the elimination of the bad data.

6.1.5 The Configuration of Power System Networks

An incorrect network topology is one of the sources for an unreliable state estimation [7, 39]. The power system state estimator is based on network configurations that represent the true status of the system and is assumed to be correct. However, due to events that produce incorrect topology, the state estimation solution may not converge or yield grossly inaccurate states. The error in topology may stem from (1) incorrect statuses of circuit breakers and switching device, which may be caused by loss of telemetry, or (2) error in the primitive network model [7].

On the other hand, distribution systems have largely extended and converted from being passive systems that are modelled only as load to be active networks with distributed generation (DG). Furthermore, the availability of advanced sensors and meters in modern distribution systems along with the expansion of distribution systems has led to the development of the distribution management system (DMS) with similar functionality as the EMS [10, 38]. In a similar manner to that of the EMS, the monitoring and control decisions of the DMS are supported by the Distribution Automation System (DAS) which offers connections with substations, feeders, circuit breakers and capacitor banks [38]. However, the initiative of the smart grids is the main motivation to improve the applications of the state estimation of the distribution systems [10, 29, 40]. Nevertheless, the configuration of the transmission systems is different to that of the distribution systems. The transmission systems are ring-shaped networks with balanced three-phase loads and three-phase balanced loads with bulk generation units, whereas the distribution system has the following predominant characteristics [29–31]:

1. The radial configuration of the power distribution networks.
2. The high *R*/*X* ratio of the power feeders.
3. Limited power measurements (low redundancy).
4. Unbalance loading with existing single- and double-phase loads.
5. Sparse and uneven incorporation of DGs (different types and sizes of DGs are used largely with the presence of many traditional passive distribution systems). However, there is no conventional bulk generation at the distribution side.

The above characteristics are challenging for many power applications such as load-flow studies and the coordination of protective equipment. Nevertheless, there is still a research gap as the early state estimation studies were applied to the transmission systems. The studies of state estimation of distribution systems have just been published in the 1990s [32, 33]. Most literature of distribution system state estimation adopts techniques that were developed for load flow of radial systems to the state estimation despite that the measurement Jacobian matrix of the state estimator is different to that of the load-flow studies (such as Newton-Raphson) [34, 35]. Therefore, research gaps exist regarding the measurement paucity, the radial configuration and the effect of *R*/*X* ratio on the performance of the state estimator. However, more details and formulation of the above issues are presented in the following chapters.

6.2 *Statistical and Data Analysis Challenges (Operator Side)*

6.2.1 Operator-Monitor Challenges at the Control Centre

The operator and the programmers must work together at the EMS and DMS, to manage the collected data and the state estimation packages to perform a reliable state estimation by implementing post-processing operations such as:

- Detecting bad data and applying protective actions against security attacks (which became a challenging task after the extension of the power systems), together with increasing cybersecurity against potential malfunctions of the communications [12, 13, 15].
- Developing new solution methods of state estimation to be used for different situations such as ill-conditioning or incomplete observability.
- Updating the historical and forecasted data that is used in the measurement set.
- Planning studies regarding the meter placement strategies (including the PMUs) based on analytical tests to the statistical nature of the available measurements. That means identifying the sources of collinear measurements and detecting the high-influential points.

- Maintenance and repairing schedules of the measuring devices and the communication means.

6.2.2 Metrics of Quality of State Estimator Solution

The assessment of the current state of the network and the accuracy of security analysis and market functions depends highly on the quality of the state estimator solution. The quality of an SE solution is typically measured using one or more of the following indices [6, 9, 36, 37]:

- Difference between the estimated and telemetry measurements (i.e. measurement residuals)
- Difference between estimated and true values (i.e. estimation residuals)
- Convergence rate of the state estimator
- The condition number that measures the numerical sensitivity of the state estimator and indicates the ill-conditioning problem, if existent

On the other hand, the values of these indices depend upon many factors that are discussed in the previous subsection such as the measurement availability, quality and redundancy and the model of the power system network. More detailed information about the operator efforts to improve the assessment criteria is explained in the next chapters.

7 The Proposed Solutions

The first stage of achieving the chapter's goals contains intensive investigations to the numerical characteristics of the state estimation within the scope of recent monitoring technologies, the latest optimisation techniques and the modern statistical analysis methodologies. The approach of the first stage is to 'stand on the shoulders of giants' to gain a better view and to review the existing methodologies of the power system state estimation. This approach is inspired by the path Fred Schweppes had followed in [4] when adopting nonlinear regression algorithms (new at that time) of statistics' giants [38, 39, 41] for solving the problem of estimation in power systems. Likewise, the analytical research of this chapter covered the tasks:

1. The factors that affect the numerical stability of the state estimator and the quality of the state estimation solution have been investigated and classified based on the nature of the problem and the available research gaps. This analysis includes revising the existing criteria for measuring the numerical stability of the state estimator and the accuracy of the estimated states. Moreover, the research gaps relevant to the problem statement were explored to facilitate applying new solution methods and assessment metrics. The different reasons behind ill-conditioning and the criteria for identifying ill-conditioned state estimators are investigated along with the feasible mathematical and technical

solutions to propose the practical solutions. Furthermore, the impact of the radial configuration of the distribution systems and the problem of meter placement (including the PMUs) are also reviewed from the state estimation perspectives in order to facilitate developing an effective meter placement algorithm.
2. The mutual effect of the methodologies on the state estimation is investigated, and a comprehensive study approach is presented. Hence, the hidden effect of the meter placement and collinear measurements and the hidden cross effect between collinearity and the high-influential measurements are revised in the contexts of the latest statistical theories.

The contribution of this chapter includes the following methodologies and techniques that have been developed by the researcher to achieve the research aims:

1. A comparative study has been implemented and published to review and evaluate the existing solution methods of state estimation. The comparative study [10] tested five solution methods (including the conventional state estimation approach) with different operating conditions to explore the response of the conventional solution methods [21, 22] in the case of ill-conditioning and the presence of high *R*/*X* ratios for the power system feeders.
2. Based on the comparative study of the state estimation solution methods, a new solution method using a regularisation algorithm is developed for enhancing the response of the state estimator, in particular, the ill-conditioned state estimator. The proposed regularisation algorithm responds to the adverse effect of the high *R*/*X* ratios and the bad convergence rate caused by the high rate of virtual measurements and solves the ill-conditioning problem without manipulating the measurement set [9, 26].

The main steps of the proposed algorithm resemble the conventional WLS state estimator by the NE approach, except steps 5–10:

Algorithm 1: Proposed regularisation method

---

1. **Initialise** state vector x^k for a flat start ($|V|= 1$ & $\theta = 0$), and **Set** tolerance τ and the iteration counter k.
2. **while** *tolerance* $> \tau$ **do**:
3. Construct the Jacobian matrix, H
4. Compute the gain matrix $G = H^T WH$
5. Calculate the condition number $\kappa(G)$
6. **if** $\kappa(G) < 0.5\times10^{12}$ **go** to step **11**; **else**
7. Do SVD.
8. Determine λ_{max}, and compute Epsilon factor.
9. Construct the regularised G matrix using the *Epsilon factor*.
10. **end if**
11. **Solve** (5) for Δx^k
12. **Update** $x^{k+1} = x^k + \Delta x^k$
13. **Set** $k \leftarrow k + 1$
14. **end while**

Alternatively, if there is a concern about the explicit ill-conditioning situation, a condition number of 10^{14} can be chosen as a decision criterion. However, a condition number higher than 10^{14} usually refers to a rank-deficient matrix. Therefore, the decision can be made without checking the condition number of the gain matrix, as the rank status can be predicted by analysing the Jacobian matrix directly.

3. Developing a new metaheuristic intelligent technique [40], which is the optimisation using the river formation dynamics (RFD) for finding the optimal placement of the PMUs. The modified version of the RFD technique is used for the first time in the power system applications [42].

   The RFD is a metaheuristic optimisation algorithm based on the swarm population approach. Although the RFD has been suggested for solving optimisation problems since 2008 [43], yet this research is the first application of the RFD in power systems studies [40, 42]. The RFD algorithm imitates how water drops can form a river channel by reducing land altitudes while moving from water sources to the destination, that is, from mountains to the sea level [44]. The sea level is the global solution to the river formation process.
4. Adopting a financial analysis technique for detecting and solving the numerical problems caused by the existing measurements or the new meters installed in the system for increasing the measurement redundancy. Particularly, the problem of multicollinearity among the measurements affects the numerical stability and even the solvability of the state estimator. A modified version of a statistical method which resembles that of financial inflation analysis is proposed to select the locations for the PMUs to be installed in without creating new collinearities. The incremental meter placement algorithm is based on the so-called variance-decomposition proportions to identify the candidate list of the host buses.
5. Developing a comprehensive study that combines the conventional meter and PMUs in a unified meter placement algorithm and provides a solution to both the multicollinearity and the high-leverage points simultaneously. The proposed methodology also uses the cuckoo search optimisation for selecting the optimal locations of the incremental meters. Statistical studies [45–48] have achieved crucial steps on identifying the high-leverage points and the sources of the multicollinearity problem. In this context, several authors [46–48] proved that the HLPs are the main source for multicollinearity and defined a group of observations, which is named as the leverage collinearity-influential observations. The collinearity-influential observations can change the multicollinearity pattern by creating or hiding collinear measurements in the measurement set.

8 Conclusion

This chapter presents the mathematical formulation of the problems relevant to the power system state estimation and reviews the fundamental functions of the state estimation. The assessment criteria of the accuracy and numerical stability of the state estimator are also presented with the factors that affect the performance of the state estimation. Although the conventional iterative WLS state estimation is numerically sensitive and likely unstable, it has a relatively simple solution algorithm and takes the measurement errors into consideration, unlike the linear state estimator. Therefore, the alternative solution methods can be classified as—an intrinsic regularisation approach, a PMU-based methodology and hybrid analytical-technical methods that are based on meter placement strategies with statistical analysis to the characteristics of the measurements and the state estimator. However, the recommended approach is a comprehensive method that combines the efforts of the utilities, the power system operator and the programmers.

References

1. Anderson, C. W., Santos, J. R., & Haimes, Y. Y. (2007). A risk-based input–output methodology for measuring the effects of the August 2003 northeast blackout. *Economic Systems Research, 19*, 183–204.
2. Grainger, J. J., & Stevenson, W. D. (1994). *Power system analysis* (Vol. 621). New York: McGraw-Hill.
3. Abur, A., & Exposito, A. G. (2004). *Power system state estimation theory and implementation*. Boca Raton: CRC Press.
4. Schweppe, F. C. (1970). Power system static state estimator-parts I through III. *IEEE Transactions on Power Apparatus and Systems, PAS-89*, 120–135.
5. Monticelli, A. (2000). Electric power system state estimation. *Proceedings of the IEEE, 88*, 262–282.
6. Wu, F. F., & Monticelli, A. (1985). Network observability: Theory. *IEEE Transactions on Power Apparatus and Systems, PAS-104*, 1042–1048.
7. Liu, J., Ponci, F., Monti, A., Muscas, C., Pegoraro, P. A., & Sulis, S. (2014, May). Optimal meter placement for robust measurement systems in active distribution grids. *IEEE Transactions on Instrumentation and Measurement, 63*, 1096–1105.
8. Abood, H. G., Sreeram, V., & Mishra, Y. (2016). Incremental placement of PMUs for enhancing state estimation accuracy. In *2016 Australasian Universities Power Engineering Conference (AUPEC)* (pp. 1–6).
9. Abood, H. G., Sreeram, V., & Mishra, Y. (2019). A new algorithm for improving the numerical stability of power system state estimation. *IEEJ Transactions on Electrical and Electronic Engineering, 14*, 358–365.
10. Abood, H. G., Sreeram, V., Al-Saadi, H., & Mishra, Y. (2017). Solution methods of ill-conditioned power system state estimation: A comparative study. In *2017 IEEE region 10 symposium (TENSYMP)* (pp. 1–6). IEEE.
11. Chen, J., & Abur, A. (2006). Placement of PMUs to enable bad data detection in state estimation. *IEEE Transactions on Power Systems, 21*, 1608–1615.

12. Mohammadpourfard, M., Sami, A., & Weng, Y. (2017). Identification of false data injection attacks with considering the impact of wind generation and topology reconfigurations. *IEEE Transactions on Sustainable Energy, 9*(3), 1349–1364.
13. Majumdar, A., & Pal, B. C. (2016). Bad data detection in the context of leverage point attacks in modern power networks. *IEEE Transactions on Smart Grid, PP*, 1–1.
14. Mohammadpourfard, M., Sami, A., & Seifi, A. R. (2017). A statistical unsupervised method against false data injection attacks: A visualization-based approach. *Expert Systems with Applications, 84*, 242–261.
15. Abood, H. G., Sreeram, V., & Mishra, Y., (2020). An incremental meter placement method for state estimation considering collinear measurements and High Leverage Points, 13(1).
16. Phadke, A. G. (2002). Synchronized phasor measurements-a historical overview. In *Transmission and distribution conference and exhibition 2002: Asia Pacific* (pp. 476–479). IEEE/PES.
17. Phadke, A. G., Thorp, J. S., Nuqui, R. F., & Zhou, M. (2009). Recent developments in state estimation with phasor measurements. In *2009 Power systems conference and exposition, PSCE'09* (pp. 1–7). IEEE/PES.
18. De La Ree, J., Centeno, V., Thorp, J. S., & Phadke, A. G. (2010). Synchronized phasor measurement applications in power systems. *IEEE Transactions on Smart Grid, 1*, 20–27.
19. Cook, R. D., & Weisberg, S. (1982). *Residuals and influence in regression*. New York: Chapman and Hall.
20. Esmaili, M., Gharani, K., & Shayanfar, H. A. (2013). Redundant observability PMU placement in the presence of flow measurements considering contingencies. *IEEE Transactions on Power Systems, 28*, 3765–3773.
21. Manousakis, N. M., Korres, G. N., & Georgilakis, P. S. (2012). Taxonomy of PMU placement methodologies. *IEEE Transactions on Power Systems, 27*, 1070–1077.
22. Abood, H. G., & Sreeram, V. (2014). A review on phasor measurement units placement for state estimation studies. In *2014 Australasian Universities Power Engineering Conference (AUPEC)* (pp. 1–6).
23. Nazari-Heris, M., & Mohammadi-Ivatloo, B. (2015). Application of heuristic algorithms to optimal PMU placement in electric power systems: An updated review. *Renewable and Sustainable Energy Reviews, 50*, 214–228.
24. Madani, V., Parashar, M., Giri, J., Durbha, S., Rahmatian, F., Day, D., et al. (2011, March 20–23). PMU placement considerations — A roadmap for optimal PMU placement. In *2011 IEEE/PES Power Systems Conference and Exposition (PSCE)* (pp. 1–7). Phoenix.
25. Madtharad, C., Premrudeepreechacharn, S., & Watson, N. R. (2003). Power system state estimation using singular value decomposition. *Electric Power Systems Research, 67*, 99–107.
26. Abood, H. G., Salman, G. A., & Kaittan, A. S. (2019). A regularized approach for solving ill-conditioned state estimation of distribution systems. *Elektrotehniški Vestnik, 86*(3), 137–143.
27. Nuqui, R. F., & Phadke, A. G. (2007). Hybrid linear state estimation utilizing synchronized phasor measurements. In *2007 IEEE Lausanne Power Tech* (pp. 1665–1669).
28. Chakrabarti, S., Kyriakides, E., Ledwich, G., & Ghosh, A. (2010). Inclusion of PMU current phasor measurements in a power system state estimator. *IET Generation, Transmission & Distribution, 4*, 1104–1115.
29. Jones, K. D., Thorp, J. S., & Gardner, R. M. (2013). Three-phase linear state estimation using phasor measurements. In *2013 IEEE Power & Energy Society general meeting* (pp. 1–5).
30. Haughton, D. A., & Heydt, A. (2013). Linear state estimation formulation for smart distribution systems. *IEEE Transactions on Power Systems, 28*, 1187–1195.
31. Korres, G. N., & Manousakis, N. M. (2012). A state estimation algorithm for monitoring topology changes in distribution systems. In *2012 Power and Energy Society general meeting* (pp. 1–8). IEEE.
32. Meliopoulos, A. S., & Zhang, F. (1996). Multiphase power flow and state estimation for power distribution systems. *IEEE Transactions on Power Systems, 11*, 939–946.
33. Ke, L. (1996). State estimation for power distribution system and measurement impacts. *IEEE Transactions on Power Systems, 11*, 911–916.

34. Eminoglu, U., & Hocaoglu, M. H. (2005). A new power flow method for radial distribution systems including voltage dependent load models. *Electric Power Systems Research, 76*, 106–114.
35. Jiang, W., & Vittal, V. (2006). Optimal placement of phasor measurements for the enhancement of state estimation. In *2006 IEEE PES power systems conference and exposition, PSCE '06* (pp. 1550–1555). Atlanta, Oct. 29 2006–Nov. 1 2006.
36. Zhu, K., Nordstorm, L., & Ekstam, L. (2009, July 26–30). Application and analysis of optimum PMU placement methods with application to state estimation accuracy. In *2009 Power & Energy Society general meeting, PES '09* (pp. 1–7). Calgary: IEEE.
37. Ebrahimian, R., & Baldick, R. (2001). State estimator condition number analysis. *IEEE Transactions on Power Systems, 16*, 273–279.
38. Anderson, T. W. (1951). Estimating linear restrictions on regression coefficients for multivariate normal distributions. *The Annals of Mathematical Statistics, 22*, 327–351.
39. Dopazo, J. F., Klitin, A., Stagg, G. W., & Slyck, L. S. V. (1970). State calculation of power systems from line flow measurements. *IEEE Transactions on Power Apparatus and Systems, PAS-89*, 1698–1708.
40. Abood, H. G., Sreeram, V., & Mishra, Y. (2016). Optimal placement of PMUs using river formation dynamics (RFD). In *2016 IEEE international conference on Power System Technology (POWERCON)* (pp. 1–6).
41. Stagg, G. W., & El-Abiad, A. H. (1968). *Computer methods in power system analysis*. New York: McGraw-Hill.
42. Rabanal, P., Rodríguez, I., & Rubio, F. (2017). Applications of river formation dynamics. *Journal of Computational Science, 22*, 26–35.
43. Rabanal, P., Rodriguez, I., & Rubio, F. (2008). Solving dynamic TSP by using river formation dynamics. In *2008 Fourth international conference on Natural Computation, ICNC '08* (pp. 246–250).
44. Rabanal, P., Rodríguez, I., & Rubio, F. (2013). An ACO-RFD hybrid method to solve NP-complete problems. *Frontiers of Computer Science, 7*, 729–744.
45. Benedito, R. A. S., Alberto, L. F. C., Bretas, N. G., & London, J. B. A. (2014). Power system state estimation: Undetectable bad data. *International Transactions on Electrical Energy Systems, 24*, 91–107.
46. Bagheri, A., Habshah, M., & Imon, R. (2012). A novel collinearity-influential observation diagnostic measure based on a group deletion approach. *Communications in Statistics-Simulation and Computation, 41*, 1379–1396.
47. Bagheri, A., & Midi, H. (2012). On the performance of the measure for diagnosing multiple high leverage collinearity-reducing observations. *Mathematical Problems in Engineering, 2012*, 1–16.
48. Midi, H., & Mohammed, A. M. (2015). The identification of good and bad high leverage points in multiple linear regression model. *Mathematical Methods and System in Science and Engineering*. 147–153. http://www.wseas.org/main/books/2015/Tenerife/MATH.pdf.

Part II
Smart Mobility

A Vehicular Internet of Things (IoT) System for High-Granularity Air Quality Monitoring in Smart Cities

Kavi Kumar Khedo, Prashant Shenoy, Khadidjah Zeadally, Khemraj Awotar, and Lamesh Ramdani

1 Introduction

As per the Global Health Observatory (GHO) data (https://www.who.int/gho/phe/outdoor_air_pollution/burden_text/en/), in the year 2016, ambient air pollution was responsible for 4.2 million deaths. Worldwide, ambient air pollution is estimated to cause about 16% of the lung cancer deaths, 25% of chronic obstructive pulmonary disease (COPD) deaths, about 17% of ischaemic heart disease and stroke and about 26% of respiratory infection deaths. Indeed, particulate matter pollution is a serious environmental health problem that affects people worldwide, but low- and middle-income countries experience this burden the most.

Internet of Things (IoT) is the concept of interrelated devices with unique identifiers that transmit data over a network with minimum human interactions [1]. With the latest developments in sensor technologies, IoT systems have great potential to be used as effective solutions for air quality monitoring on a real-time basis [2]. Use of appropriate gaseous sensors with wireless communication capabilities can provide real-time ambient air quality information to the users.

Existing air quality monitoring systems worldwide are mostly stationary stations providing air quality information to environmental authorities and other social partners. These systems are very useful in keeping people aware of the quality of the air surrounding them, eventually helping them to cater for their health and take necessary precautions needed. However, as they are fixed in specific locations,

K. K. Khedo (✉) · K. Zeadally · K. Awotar · L. Ramdani
Department of Digital Technologies, FoICDT, University of Mauritius, Réduit, Mauritius
e-mail: k.khedo@uom.ac.mu

P. Shenoy
College of Information and Computer Sciences, University of Massachusetts Amherst, Amherst, MA, USA

M. M. Banat, S. Paiva (eds.), *Smart Technologies for Smart Cities*, EAI/Springer Innovations in Communication and Computing,
https://doi.org/10.1007/978-3-030-39986-3_6

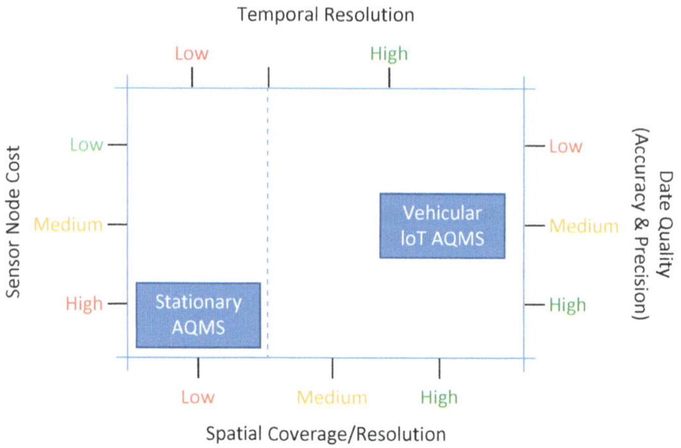

Fig. 1 Spatial and temporal resolution of AQMS

the spatial coverage of such systems for proper monitoring of air quality in urban environment is still a challenge. Air quality information are therefore available for specific locations only, and stakeholders are not able to get air quality information for the regions that are not covered by the fixed air quality stations. Moreover, it will be very expensive to replicate the fixed stations to all regions of the island. One possible solution to overcome this problem is to use vehicular sensor networks for the air quality monitoring. Figure 1 illustrates the spatial and temporal resolutions of both a stationary AQMS and vehicular IoT AQMS. As shown in Fig. 1, stationary AQMS have high data quality; however they also have a high cost and low spatial coverage. On the other hand, vehicular IoT AQMS have medium data quality with medium cost and high spatial coverage. Vehicular IoT AQMS is therefore an interesting solution for air quality monitoring in large geographical areas.

Hence this work aims at developing a low-cost IoT module that may be mounted on vehicles forming a vehicular sensor network for real-time monitoring of air quality. As the vehicles, such as buses, will be travelling in different locations at regular interval of time, fine-grained air quality information can be collected in the different locations, thus providing better spatial coverage than stationary air quality monitoring stations. Real-time information about the state of air quality including the concentrations of the two major pollutants, particulate matter $PM_{2.5}$ and PM_{10}, that are measured on an hourly basis in target regions are obtained and stored on a remote server where a number of air quality data analytics are performed. Based on the results of the analysis performed, necessary alerts are generated and communicated through a mobile application for abnormally high level of air pollution detected in a specific area.

A mobile application has been developed to allow visualization of the air quality information. Through the mobile application, the general public is informed about the quality of the air in different regions. The air quality data are presented visually using different coding schemes and format for easy interpretation and understanding. People may use the air quality mobile application to better plan their travel and take necessary precautionary measures.

In this chapter, related works are critically analysed in Sect. 2, and in Sect. 3 a domain analysis of air quality monitoring in Mauritius is carried out. Section 4 assesses the different technologies available for the development of an air quality monitoring system, and Sect. 5 identifies the requirements for the low-cost vehicular IoT air quality monitoring system. The design of the proposed vehicular IoT system for air quality monitoring is given in Sect. 6, and Sect. 7 discusses the results.

2 Related Works

Due to the latest technological advancement in micro-electromechanical systems (MEMS) and chemical microsensors, the use of tiny and inexpensive sensors to monitor various physical phenomena has become technically and economically feasible. Wireless sensor network (WSN) consists of many sensor nodes scattered in the physical environment to measure a particular phenomenon. They have the capabilities to measure at a fine-grained level different kinds of parameters such as humidity, temperature, pH and also gases in the atmosphere with a reasonable level of accuracy. Commercially calibrated discrete gas sensors for sensing concentration of gases like CO2, NO2, CO and O2 are now readily available. These pre-calibrated gas sensors are integrated with wireless sensor motes to form a monitoring station for field deployment. In this section some related works on the use of sensor and IoT technologies for air quality monitoring are described.

A team at the SASTRA University in India implemented an IoT air quality monitoring system (AQMS) for both indoor and outdoor air quality monitoring in real time [3]. The system used Raspberry Pi microcontroller as a processing unit to design sensor web nodes consisting of gas sensors, and Wi-Fi is used as transceiver for data communication between the web node and a server. When nodes are deployed, the obtained data from each node are archived firstly in a local database and then are sent to the ThingSpeak cloud database. A cloud service was used in the system in order to execute a number of operations such as data management, data-centric aggregation, decision-making and data analytics for the data from the multiple web node, which normally requires more computing resources than available at the node level. Moreover, a mobile application was developed in order to allow users to access the data from the cloud database. The users can monitor the real-time data streams by logging to the public channel of the ThingSpeak cloud services. While this work proposed an interesting design architecture for real-time air quality monitoring, there is no evidence of a wide-scale deployment, and additionally, since the monitoring nodes are static, it will require large number of nodes for a high-resolution spatial coverage.

Air Monitors proposed a wireless air quality compact system called AQMesh (https://www.aqmesh.com/) for outdoor gas and particle monitoring. AQMesh provides near real-time localized air quality information and data analysis. It consists of wireless networked array of electrochemical sensors monitoring up to five pollutant gases with built-in wireless mobile GPRS technology and storage on the secure server. Air quality readings are provided to the users in meaningful patterns, trends, tables and graphs and diagnosis deduced by algorithms. AQMesh is considered as an air quality monitor that can deliver localized, real-time readings, aiming to improve the scope and accuracy of gathering air quality data. Recent co-location comparison trials against certified reference equipment continue to prove AQMesh performance and reliability for localized air quality monitoring [4]. While AQMesh demonstrates a very high level of accuracy for air quality monitoring, it is not a highly scalable solution for implementation in large cities and is only appropriate for localized areas.

The SensorWebBike [5, 6] is a Geospatial Information Infrastructure designed to support a participative sensing approach for urban environmental monitoring. Volunteer bikers are used as citizen-sensors who measure environmental parameters, by means of a small sensor box mounted on their bikes. The SensorWebBike system uses the concept of participatory sensing with the aim to improve the scientific understanding of monitored spatial phenomena, which cannot be fully detected by static air quality stations. A circuit board was developed called ‘AirQuino’, Arduino shield compatible, integrated with low-cost and high-resolution sensors, dedicated to the monitoring of environment and air quality, road pavement quality (accelerometer) and the indices of well-being in an urban environment. The board integrates a microprocessor unit that acquires all the sensor readings and analyses fast data from accelerometer and noise sensor. Through General Packet Radio Service (GPRS) technology, the sensor transmits geo-located data on the environment and air quality to a server connected to the applications and web server allowing to visualize real-time observations on a web browser. The components are organized in typical client-server architecture and interact from the sensing process to the representation of the results to the end users. Through common web or mobile browsers, all the collected data are visualized in table or chart format or tracks and spot values on a Google mashup. Figure 2 illustrates the SensorWebBike framework components.

The AQM 65 Ambient Air Monitoring System (https://www.aeroqual.com/product/aqm-65-air-quality-monitoring-station) is a fully integrated air monitoring station that delivers ‘near reference’ levels of performance. It is of size of a large

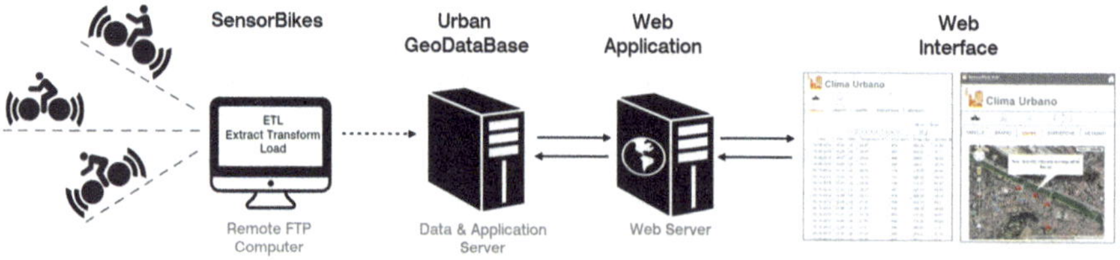

Fig. 2 The SensorWebBike framework components [5]

suitcase, and it can measure up to 20 different gaseous and particulate pollutants and environmental parameters simultaneously. AQM 65 forms a lower-cost, high-density ambient pollution monitoring system for an industrial perimeter. The AQM 65 ensures air quality data is reliable and robust and traceable back to recognized international standards, e.g. USEPA (40 CFR Part 53) and EU (2008/50/EC). The unit is easily transported by car, carried by two people, and installed and operational in under 30 minutes. This makes deployment and relocation quick and easy. It also opens up monitoring locations that were previously inaccessible to traditional monitoring stations. Because the enclosure is small, the energy required for temperature control is much lower than for a traditional monitoring station. Lower-power consumption opens up the possibility of remote power supply and decreases the total cost of ownership.

Arroyo et al. [7] proposed an air quality monitoring that is composed of a distributed sensor network connected to a cloud system forming a wireless sensor network (WSN). The monitoring nodes are implemented using low-power ZigBee motes, and the nodes transmit air quality measurement data to a cloud system through a gateway. Data from the sensor network are stored, processed and visualized at the cloud computing platform. Figure 3 illustrates the sensor network architecture used for implementation of the proposed system. The researcher used artificial intelligence techniques to better perform detection of compounds and contaminants from the data. It is claimed that the proposed system is a low-cost and low-power consumption which has the potential to efficiently measure air quality in large scale since the monitoring nodes can be easily replicated and deployed. However, it is observed that tests with respect to the applicability of the proposed system conducted by Arroyo et al. [7] for the detection of common volatile organic compounds were limited to laboratory environment only. Moreover, the system has not been deployed on a large scale and real conditions testing have not been performed. Thus the applicability of the system for the fine-grained large-scale monitoring of air quality in a city cannot be established.

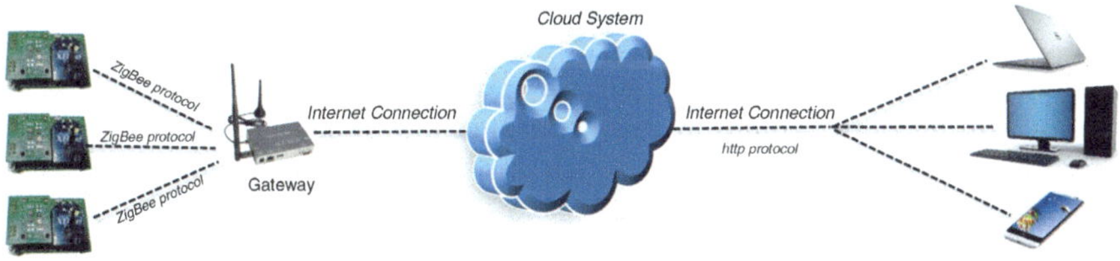

Fig. 3 Sensor network architecture [7]

3 Domain Analysis: The Case of Mauritius

In Mauritius, sustained economic development and changes in consumption and production patterns have significantly contributed to the release of different pollutants in the atmosphere over the years, thereby adversely affecting air quality. The three main sources of air pollution in Mauritius are from the burning of fossil fuel for heat generation in industries, electricity generation and transportation among others like sugarcane burning. As per Statistics Mauritius [8], the greenhouse gas (GHG) emissions went up by about 3% from 2016 to 2017. Gross emissions increased from 5403 to 5572 thousand tonnes of CO2 equivalent. This was mainly due to a rise in emission from the energy industries which accounted for the largest share of emissions (76%) in 2017. The transport industries accounted for 25% and the manufacturing industries 8% of the carbon dioxide (CO_2) emission.

Several initiatives to reduce air pollution have been introduced by the Government of Mauritius such as policies to reduce emissions of harmful gases from industries and transport. But unfortunately air pollution is still posing a threat to the island since the economic success of Mauritius has led to a significant increase in road transport and industrial activities resulting in more air pollution with smoke emission and other pollutants. These pollutants are increasingly causing harm to the health of the Mauritian people, and there has been substantial increase in respiratory and cardiovascular-related diseases. Over the last few years, an increase of 21% in air pollution has been registered by the statistical office in Mauritius [8].

Indeed, as Mauritius is becoming more industrialized, urbanized and densely populated nation, maintaining a good ambient air quality becomes a major challenge. The Ministry of Environment has somehow tried to adopt some measures to improve air quality like reducing emissions at sources, keeping industrial waste away from residential areas and monitoring the ambient air of suspected locations. AQM has been recently introduced in Mauritius by the Ministry of Environment. The National Environmental Laboratory (NEL) of the Department of Environment has installed two ambient air quality monitoring stations, one at the Mauritius Meteorological Services (MMS), Vacoas, and the other at Islamic Cultural Centre (ICC), Port Louis.

These stations are in operation since 20 August 2015. The purpose of these stations is to regularly monitor particulate matter (PM_{10} and $PM_{2.5}$) and to study the trends of air quality around the island. If any sign of degradation in air quality is identified, the National Environmental Laboratory (NEL) initiates remedial actions to protect the health of citizens. The automatic analysers and equipment at the stations measure the concentrations of major pollutants such as sulphur dioxide, oxides of nitrogen, carbon monoxide, ozone, particulate lead and respiratory suspended particles (PM_{10}). Data obtained from these stations are sent to scientific officers of the NEL for analysis. The NEL checks whether the data acquired is compliant to the particulate matter levels with the approved limits of the Ambient Air Quality Standards recommended under the Environment Protection Act 2008 or not [9].

However, the network of fixed ambient air monitoring stations suffers from a number of major shortcomings. Firstly, the cost of acquisition and operating the monitoring stations are extremely high, thus limiting its scalability. Secondly, the system is highly localized resulting in spatial gaps in the measurements of the air quality around the island. Given the limitations of the current air quality monitoring system in Mauritius, there is a need for the development of a low-cost, highly scalable and high-granularity system.

Besides, it has been decided by the Government that 11 smart cities are going to be constructed in Mauritius. This will contribute to the modernization of Mauritius and will help in fulfilling the vision of a smart Mauritius island. One of the investment opportunities is to improve the ICT infrastructure. This concept deals with big data analytics for the efficient functioning and real-time management of smart cities. The ICT technologies are expected to offer the residents with a better traffic control system, free-space parking detection, security, healthcare, improved and cost-effective power supply and management and most importantly proper AQM system.

There is no definite air quality index used in Mauritius. There is the Ambient Air Quality and Emission Standards that set the guidelines for the maximum limits for each pollutant as well as measurement methods for the pollutants [9]. The Ministry of Environment is currently developing a new set of air quality and emission standards to follow the Environment Protection (National Standards for Ambient Air Quality and Stack Emission) Regulations 2011. The Environment Protection Regulations 2011 instructs to calculate the AQI using breakpoint and standard formulas.

4 Technological Analysis

A vehicular sensor network system for monitoring of air quality should consist of several components including the monitoring module, data transmission, data storage and data analytics. There is a myriad of IoT technologies available to support the development and implementation of such a system. In this chapter a technological analysis is carried out, and the different technological options are considered for the vehicular sensor network, and then the requirements for the air quality monitoring system are derived.

4.1 Microcontrollers

The microcontroller can be considered as the brain of the IoT system. Discrete microcontroller IC can be used in order to process the information such as the AVR family from Atmel and PIC family from Microchip. However, using the discrete microcontroller IC is more challenging as the different electronic components need to be assembled from scratch. Moreover, a PCB (printed circuit board) will be needed in order to make stable connections to each component. Boards from the Arduino or Raspberry Pi family can also be used in order to implement the monitoring module. They usually feature a microcontroller from Atmel's AVR family with a friendly prototyping board.

4.2 Communications

In order to implement a wireless sensor network, wireless communications must be established from monitoring nodes to base stations and from the base stations to a remote server. Wireless technologies should be carefully chosen for the vehicular sensor network since the monitoring nodes may be moving at a high speed (e.g. they may be placed on buses or the Metro), and data must be sent in very limited time. Therefore the latency of the wireless communication technology should be minimum. Therefore, the technology chosen must feature a high bit rate and also a decent range for communications. For the short-distance communication from monitoring nodes to base stations, low-power technologies such as BLE (Bluetooth Low Energy) and ZigBee were considered. As for the long-distance communication, technologies such as LoRaWAN and Narrowband IoT (NB-IoT) will be considered.

4.3 Power Source

In order to power the monitoring device of the vehicular sensor network, an efficient and low-cost power source needs to be considered. Two possible power sources for the air quality monitoring device are the vehicle's battery or the solar power. Each power source has advantages and limitations that should be considered. The vehicle's battery is typically rated at 12 V, and therefore, it must be converted to 5 V since most microcontrollers and microcontroller board work safely with 5 V. This can be achieved using a linear voltage regulator such as the LM7805. However, linear voltage regulators work by converting the excess voltage into heat. Therefore, linear voltage regulators are not very efficient. Another method to step down the 12 V battery voltage to a safe 5 V is by using a buck-converter switched-mode power supply (SMPS). The buck converters have efficiency of 80%. Energy from the sun can be converted into a DC voltage using solar panels. The voltage received from the solar panel depends on the power rating of the panel. In order to store the electrical energy received from the solar panel, a rechargeable battery must be connected to it, and then microcontroller will take its power from the battery. The voltage from the solar panel may be too little or too high for the battery; therefore buck-boost converter can be used. The buck-boost converter is a type of SMPS. If the voltage in its input is higher than the expected output voltage, the voltage will be stepped down, and if the voltage in its input is lower than the expected output voltage, the voltage will be stepped up.

4.4 Cloud Server

The data from the base station must be sent to a cloud server for permanent storage and necessary analysis. There are several cloud service providers for IoT systems, and the services vary from simple storage of sensor data to advance analytics. Amazon Web Service (AWS) IoT Platform is a fully managed cloud platform that allows the connection of various devices easily and securely with cloud applications. AWS IoT Platform allows sensor devices to connect to AWS services and other devices, provide secure data and interactions, process and act upon device data and enable applications to interact with devices even when they are offline. ThingSpeak is an IoT analytics platform service that allows the aggregation, visualization and analysis of live data streams from sensors in the cloud. ThingSpeak has integrated support from the numerical computing software MATLAB, allowing users to analyse and visualize uploaded data easily using MATLAB. The IBM Watson IoT Platform is a fully managed, cloud-hosted service designed to quickly make use of IoT devices. It uses the industry-standard MQTT protocol to connect devices and applications. MQTT is designed for efficient exchange of data to and from devices in real time. It allows IoT system to securely receive data from and send commands to devices. MQTT with TLS is used to secure all communication between devices and the platform service. Moreover, in the IBM Watson IoT Platform apart from having access to real-time data coming from devices, data can be stored for a period of choice, allowing access to both historical and real-time data for devices.

5 Requirements Analysis

As discussed in the Domain Analysis, there are only two fixed ambient air quality monitoring systems in Mauritius found at Vacoas and Port Louis. The limitations of these two stationary air quality monitoring systems have already been discussed. It is argued that a vehicular IoT air quality monitoring system (AQMS) will be more beneficial in terms of cost, collection of data including geographic coverage, analytical capabilities and providing feedbacks and real-time information to the population.

5.1 Air Quality Index for Mauritius

A simple AQI must be developed and tested for effective evaluation of the air quality status using the data received from the sensors. It will be inspired by the Ambient Air Quality and Emission Standards for Mauritius under the Environment Protection Act 2008 [9]. According to USEPA [10], the concentrations of the two major pollutants, particulate matter $PM_{2.5}$ and PM_{10}, must be measured on an hourly basis to calculate the AQI values for each pollutant with a specific linear formula.

5.2 Data Storage

Air quality data will be collected on an hourly basis, and a proper data storage facility should be available so as to store the real-time and historical data from the sensors. Using one of the mentioned cloud services in technological analysis will offer several advantages in terms of data storage and data management. For instance, many devices such as mobile phones can be connected to the cloud service, and the public may access the air quality data directly from the cloud. Moreover, using a cloud server entails less risk of losing data, since most cloud services provide a disaster recovery option with automatic backups of files and a good level of security for the data files.

5.3 Data Analytics

After data has been collected on the cloud, data analytics should be performed in order to generate appropriate air quality graphs and trends. Graphs can be generated to compare the air pollution in different locations or at different times and identify trends. The graphs will allow users with no knowledge about air quality to have an understanding about the air pollution in a particular location and understand how the air quality is varying hourly and daily for different locations. Based on historical air quality data and meteorological conditions, forecasting for the next days can also be done on the cloud, providing data to people to better plan their activities in order to reduce exposure to air pollution.

5.4 Data Visualization and Mobile Application

In order to visualize the air quality data and to give access to the population about information related to air quality in different locations in a real-time manner, a mobile application is proposed which will connect directly to the cloud server. It will allow users to easily understand the quality of air surrounding them and, in different regions, to take safety precautions and to keep away from unhealthy environments. It must include a user-friendly interface with visually appealing meaningful graphs to facilitate the communication of discovered trends and forecasts in air quality to various types of users, even those with low or no knowledge about air quality.

5.5 Spatial and Temporal Coverage

Sensor coverage is a well-established problem in sensor networks. Most of the work is focused on optimizing coverage in stationary networks or by controlling the movement of mobile nodes (e.g. robots) in order to maximize their coverage. The proposed system is an alternative to expensive air quality monitoring stations and should provide fine-grained air quality data for better spatial and temporal monitoring. In a vehicular IoT AQMS, the frequency and location at which the device will collect air quality data is very much determined by the speed at which the vehicle is moving. A vehicle moving at a high speed will have lower time intervals between records so that a location is not missed. On the other hand, a vehicle moving at low speed will have higher time intervals between records so as to not have redundant data for the same location. The device will need to be able to alter its time interval between records on the go.

6 System Design and Deployment

In this section, the vehicular IoT system for real-time air quality monitoring in smart cities in Mauritius is described. The main focus of the proposed work is to design a low-cost IoT module for air quality monitoring that can be easily replicated on vehicles and provide good spatial coverage of the status of air quality in a city. The vehicular-only deployment of the system is described in this chapter. Air quality sensor data are collected and transmitted to a remote server where a set of analytics tasks are performed in order to calculate the air quality index, thus providing real-time ambient air quality information to the local authorities and population. Based on the results of the analysis performed, necessary alerts are generated for abnormally high level of air pollution detected in a specific area. Moreover, a mobile application is proposed for easy visualization of the air quality data. Through the mobile application, the general public is informed about the quality of the air in different regions of a city. The air quality data are presented visually using different coding schemes and format for easy interpretation and understanding. People may use the air quality mobile application to better plan their travel and take necessary precautionary measures.

The modules developed in the vehicular IoT AQMS are the monitoring node, the base station, the cloud server and the mobile application. The monitoring node and

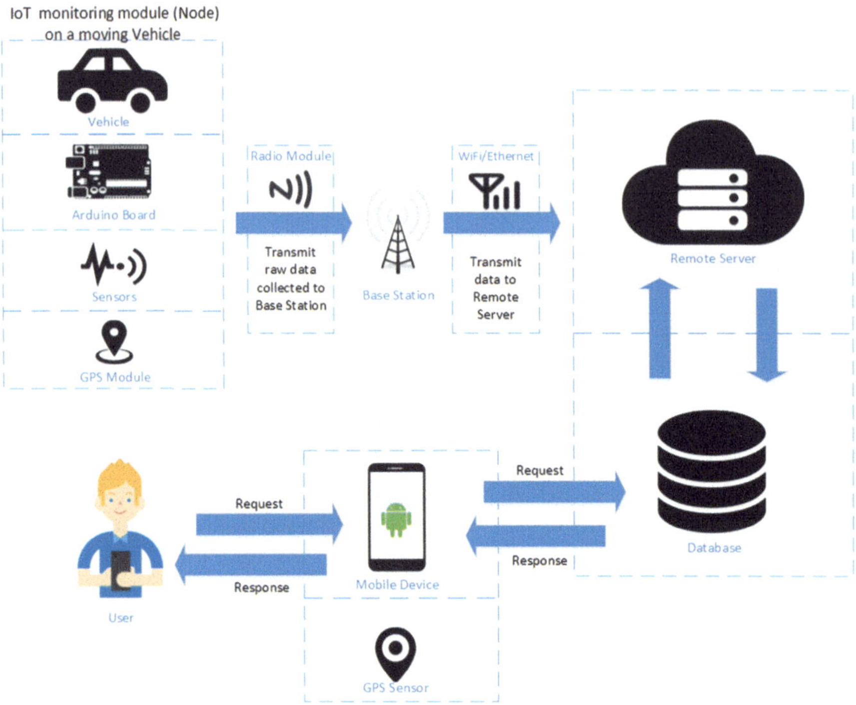

Fig. 4 Vehicular IoT AQMS sensor network architecture

the base station were developed using Arduino boards, enabling the low-cost system expected with the sensors monitoring temperature, humidity, $PM_{2.5}$ and PM_{10} and the GPS module receiving location coordinates. The cloud server was designed on ThingSpeak to allow storage, analysis and predictions using the Statistical and Machine Learning Toolbox, a feature of ThingSpeak. To meaningfully display the analysed real-time information about air quality, Android was used to build the mobile application, allowing access to code libraries online. Figure 4 illustrates the overall system architecture for the vehicular IoT AQMS.

The IoT monitoring module, also called a node, which is placed on a moving vehicle, constitutes an electronic board, a GPS module and sensors for various gases. The node gathers and stores information collected by the sensors for a limited period of time before sending to a base station via a PAN (personal area network). Nodes will be location aware through GPS allowing better spatial coverage. Once the vehicle carrying the node is close enough to a base station placed in strategic locations, the node transmits the collected data for the period.

The base station therefore receives periodic data from nodes, and it forwards these data to a cloud server using an appropriate Internet connectivity network through the GSM network, Wi-Fi or an Ethernet port depending on availability. The tasks of collecting, storing and sending data to the remote server are done in such a way so as to prevent data redundancy. Therefore, proper algorithms have been implemented in order to determine the timings and contents for the data transfer to the cloud server. For example, redundant air quality data for the same location obtained from different vehicles are not considered for transmission to the cloud server.

All the data received from the base stations are stored in the database at the cloud server, and appropriate data analytics are performed on the data in order to generate air quality reports in a real-time manner. These reports are displayed in the form of graphs to the users of the system via a mobile application. Results on air quality trends and forecast are also generated based on data collected for a long period of time. The mobile application, named AirVisual, is developed in order to communicate air quality information to users. It connects directly to the cloud server and provides real-time information to the users. AirVisual also makes use of the mobile device's GPS sensor, allowing the user to set his location and edit his preferred settings in order to have notifications of air quality whenever and at whatever frequency needed. The user is also able to view air quality information of multiple regions.

6.1 Air Quality IoT Monitoring Module

The Arduino Uno microcontroller board is used for the monitoring module. The Arduino Uno is an open-source electronics prototyping board. It provides the platform on which the sensors are connected and data is collected. The Arduino Uno is used for both the monitoring node and the base station. The IoT monitoring nodes are low-cost modules, which have no Internet connection and use ZigBee wireless communication to send its data collected to the base stations which are Internet enabled. The vehicle's 12 V outlet is used to power the monitoring node. A 12–5 V charger is connected to the 12 V outlet and a power bank is connected to the charger. The monitoring node then draws its power from the power bank. Data collected from the monitoring node are stored in arrays and then transferred to a base station using ZigBee which is a low-power WPAN technology. The module XBee is used for ZigBee communication and it is connected to the R_X and T_X pins of the Arduino. The XBee module needs 5 V of power to operate and it is supplied via the 5 V pin of the Arduino. The XBee module is placed on an XBee shield for easier interfacing.

There are two sensors that are needed in order to achieve the air quality monitoring system. The two sensors are DHT11 (for humidity and temperature) and SDS021 (for particulate matter). The sensors send analogue voltages to the Arduino which is then processed to get the data. All the sensors get their power using the 5 V of the Arduino. Figure 5 shows the connections of the IoT monitoring module to the sensors. The DHT11 sensor sends its data serially on the digital pin 2, and the SDS021 sensor sends its data through its T_X (transmit) R_X (receive) port to the Arduino.

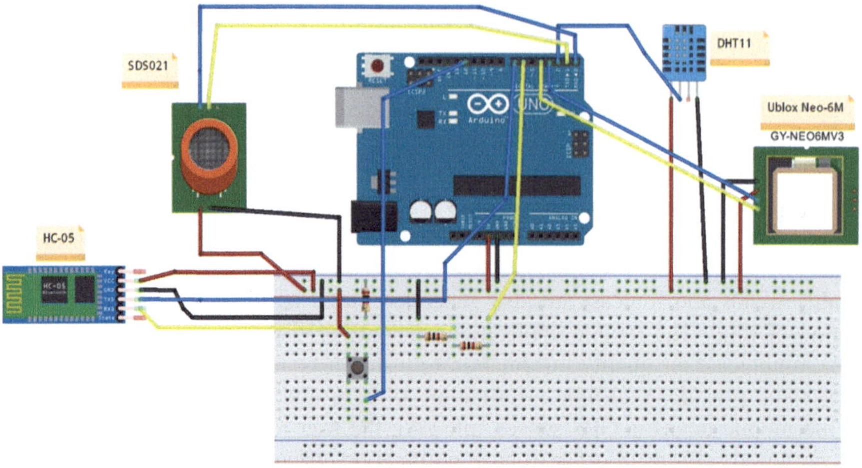

Fig. 5 IoT monitoring node to sensors connections

6.2 Base Station

For the implementation of the base station, an Arduino Yun microcontroller board is used. Similar to the monitoring node, the base station also uses the XBee board for ZigBee communication to receive data from the IoT monitoring node and is wired the same way that is connected to the T_X and R_X pins of the Arduino Yun, and it receives its power using the 5 V pin of the Arduino. Since the base station needs to additionally send data to the cloud server, the Ethernet port of the Arduino Yun is used. However, the base station is also equipped with Wi-Fi and a GSM card in cases where wired connectivity is not available.

6.3 Cloud Server

The ThingSpeak cloud service is chosen for the vehicular IoT AQMS as it is a low-cost solution and it provides the necessary data analytics features required for the system. ThingSpeak is an open IoT platform service which also acts as a remote server mainly used to store sensor data from the IoT monitoring module. The stored data are then analysed in real time using the MATLAB analytics feature. Channels are created to collect sensor data received from the IoT monitoring module, and the data are processed using MATLAB. One main channel is used to collect data from the IoT monitoring module. From this main channel, analysis is done to populate the channels created for each of the monitored locations to store hourly AQI values and then daily AQI values. Figure 6 shows the overall architecture of the ThingSpeak cloud service.

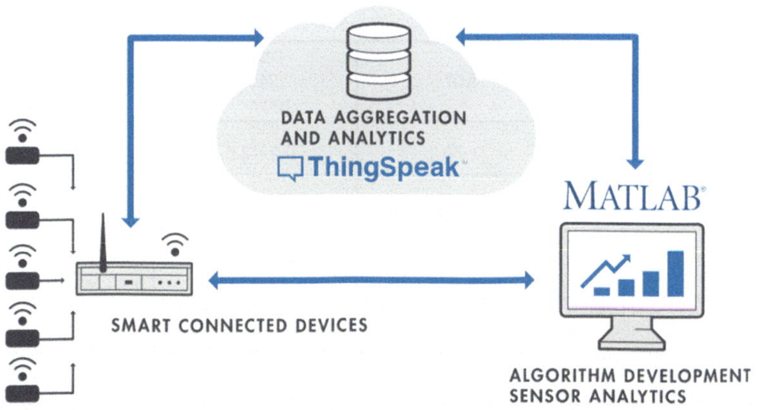

Fig. 6 ThingSpeak cloud service architecture [11]

ThingSpeak served as a storage device with data transmission available for the edge node devices but also as data exploratory analysis software. Most of the data processing are done on the remote server so that processing load on the IoT module is considerably reduced to maintain optimum power consumption. It allows the visualization, access, aggregation and then analysis of the data streams in the cloud itself and produces appropriate air quality reports to be shared with any number of users. Moreover, sensor-logging applications can be created using channels. The channel, a primary element of ThingSpeak, contains data fields, just like a regular database. It can store all data the application collects, including eight fields of any type of data along with three fields for location. The data written to the channel are then processed using MATLAB code on an hourly and daily basis to produce specific outputs such as hourly AQI values for each pollutant and for each specific location. The data are imported from the ThingSpeak database, either to a CSV file or using REST API calls with Json output format for visualization on the AirVisual mobile application. Private and public views for each channel can also be configured.

6.4 AQI Design

A simple AQI system for effective evaluation of the air quality status using the data received from the sensors is derived. It is inspired from the Environment Protection Act from the United States [10] and the Ambient Air Quality and Emission Standards for Mauritius under the Environment Protection Act 2008 [9]. Table 1 shows the AQI value breakpoints used for a given hour.

Table 1 AQI breakpoint

AQI category	AQI	$PM_{2.5}$ (24-h) (μg/m^3)	PM_{10} (24-h) (μg/m^3)
Good	0–50	0.0–12.0	0–54
Moderate	51–100	12.1–35.4	55–154
Unhealthy for sensitive groups	101–150	35.5–55.4	155–254
Unhealthy	151–200	55.5–150.4	255–354
Very unhealthy	201–300	150.5–250.4	355–424
Hazardous	301–400	250.5+	425+

To calculate the AQI for each pollutant, a linear formula is applied using the breakpoint concentrations and the AQI values that correspond to the concentration obtained as shown in Eq. 1 below:

$$\mathrm{AQI} = \left[\left\{\frac{(\mathrm{I}_{\mathrm{HI}} - I_{\mathrm{LO}})}{(B_{\mathrm{HI}} - B_{\mathrm{LO}})}\right\} * (C_{\mathrm{P}} - B_{\mathrm{LO}})\right] + I_{\mathrm{LO}} \tag{1}$$

where

B_{HI} = Breakpoint concentration greater or equal to given concentration
B_{LO} = Breakpoint concentration smaller or equal to given concentration
I_{HI} = Maximum AQI value corresponding to B
I_{LO} = Minimum AQI value corresponding to B
C_{P} = Concentration of the pollutant obtained

For example, the AQI which corresponds to an obtained $PM_{2.5}$ concentration of 32 μm/m3 is as shown in Eq. 2 below:

$$\mathrm{AQI} = \left[\left\{\frac{(100 - 51)}{(35.4 - 12.1)}\right\} * (32 - 12.1)\right] + 51 = 93 \tag{2}$$

where

$B_{\mathrm{HI}} = 35.4$
$B_{\mathrm{LO}} = 12.1$
$I_{\mathrm{HI}} = 100$
$I_{\mathrm{LO}} = 51$

Thus, the AQI for $PM_{2.5}$ will be placed in the category of moderate, instructing sensitive people to reduce heavy and prolonged exertion. AQI will be calculated for both gases.

6.5 Data Analytics and Forecasting

The sensor data (temperature, humidity, $PM_{2.5}$, PM_{10}) received from the base stations are analysed by the remote server to produce relevant air quality information that are plotted on graphs with specific parameter settings. Data needs to be read from the main channel for the last hour, and then analysed data are stored in their respective channels identified by the channel ID. Data for each field in a channel is stored in an array, for example, channel tempF contains all records from the field temperature. For the data analysis, first the data are filtered based on the location. The coordinates A_x, A_y, B_x and B_y for a specific location are retrieved from the channel Location. Once the coordinates are retrieved for a particular location, the system then iterates through the array of the data from the base station and determines which of them belong to that specific location. The sensor data for a specific location and day are sorted in the array, and the daily medians, lower and upper quartiles, are calculated. The daily values for temperature, humidity and overall AQI for every location are calculated. Furthermore, various daily trend analyses are performed such as air quality concentrations for a specific gas in a specific lap of hours or days and air quality concentrations for a specific gas in different locations. Figure 7 shows the daily trend analysis of sensor data for two different locations.

Prediction can be done by machine learning, the science of making computers learn and improve their learning over time by feeding relevant and efficient datasets [12]. The Statistics and Machine Learning Toolbox in ThingSpeak provides necessary functions for exploratory data analysis. The Classification Learner App and Regression Learner App use supervised and unsupervised machine learning algorithms to draw inferences from provided data and build predictive models. When multidimensional data analysis is done on this, it provides feature selection and other reduction methods to identify variables that impact the prediction models.

In the vehicular IoT AQMS, the dataset used is the past AQI data since 2017, retrieved from the cloud server and publicly available air quality data from epa.gov. Evidently, data from the epa.gov was chosen to match the variance of the data in Mauritius with annual average PM_{10} values being in the range of 10–30 $\mu g/m^3$. The dataset will contain the past 3 days as predictors and the present AQI as a result. To train the machine learning module, all data must scale in the range. After the machine learning training, predictions are done daily by retrieving the last three daily AQI values for every location from a channel in ThingSpeak.

Fig. 7 Daily trend analysis of sensor data for specific locations

6.6 *AirVisual Mobile Application*

A mobile application named AirVisual has been developed in order to visualize the air quality data in different locations in a real-time manner. The mobile application features a user-friendly interface with visually appealing coding scheme and meaningful graphs to facilitate the communication of discovered trends and forecasts of air quality to various types of users. Real-time hourly AQI values for each location are displayed on the map at specific locations using markers. When the markers are clicked on, all the detailed air quality information and the AQI message are displayed. When the user scrolls down, the line graphs are displayed to show how the hourly values have varied during the day. Figure 8 shows the mobile interface for visualizing the air quality data for the current location.

The AirVisual mobile application also allows users to receive notifications for air quality for a specified location and time. Users are automatically notified of the quality of the air by the AirVisual mobile application.

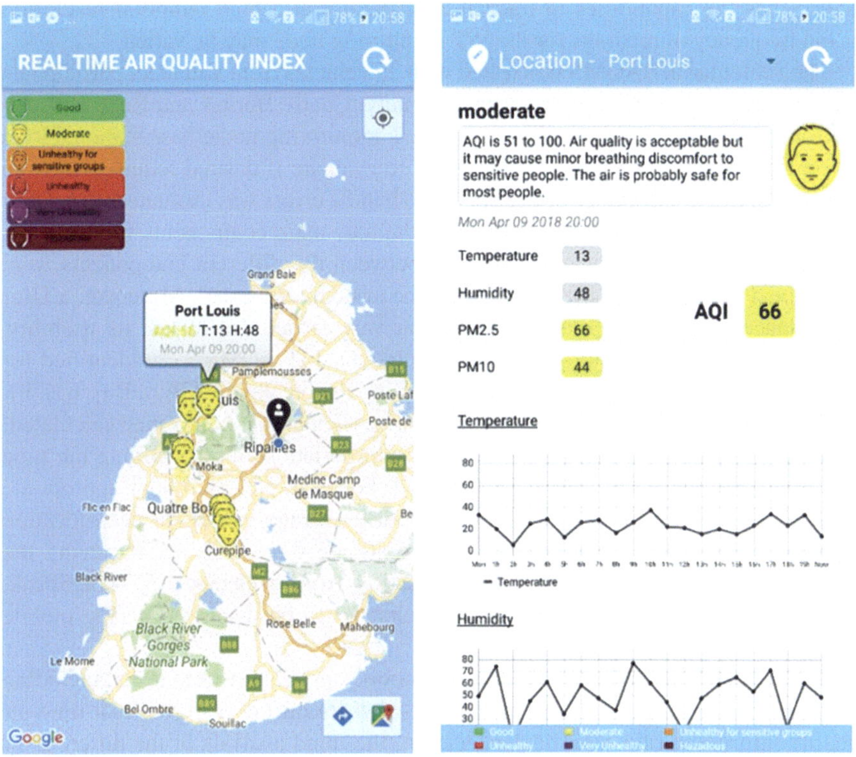

Fig. 8 Visualizing sensor data for the current location

7 Results and Discussions

The implementation of a vehicular IoT air quality monitoring system (AQMS) for smart cities is significantly more advantageous for high-granularity air quality monitoring compared to expensive fixed air monitoring stations. As discussed in the domain analysis, there are currently in Mauritius two fixed ambient air monitoring stations that have costed around 10 million rupees to be made operational, whereas the vehicular IoT AQMS cost only a few thousands rupees to implement.

Moreover, the mobility of the carrier (vehicles) enables a sensor node to cover sufficiently large geographic areas within a short period of time. Therefore higher spatial resolution of the sensed data can be achieved, and fewer number of sensor nodes are required compared to systems using stationary air quality stations. Stationary AQMS normally provides 24-hr air quality reports, thus limiting the temporal resolution of the air quality information. Air quality normally dynamically changes during a day for a region depending on the climatic conditions and other activities such as volume of traffic and industrial activities. The vehicular IoT

AQMS has the capabilities to track the air quality at a high temporal resolution as the frequency of readings for the IoT monitoring node may be varied.

The vehicular IoT AQMS was tested with 25 vehicles from volunteer participants spanning two major cities in Mauritius, namely, Quatre Bornes and Rose Hill. All the four main components of the system (IoT monitoring node, base station, cloud server and mobile application) were tested to evaluate if desired requirements are fulfilled, and also the ability of the system to handle errors was evaluated. First, unit testing was performed to check if each of the four main components was working on their own. Then, the interconnections between the different components were tested, and then data integrity throughout the system was tested. Moreover, a User Acceptance Test for the mobile application was carried out to test its usability. Finally, some areas of improvements for vehicular IoT AQMS were identified for future works. The hourly AQI is calculated using the breakpoints B_{LO}, B_{HI}, I_{LO}, I_{HI} for $PM_{2.5}$ and PM_{10}. The overall AQI for that particular location is then calculated.

Time controls were implemented to run appropriate script, specifying the time and frequency the script should be executed. For example, each hour appropriate scripts must be executed to calculate all hourly average values for each location using the time control component. Predictions are done daily by retrieving the last three daily AQI values for every location from a channel in ThingSpeak. The predicted data is stored in the same channel to be displayed on the mobile application.

The different features of the AirVisual mobile application were tested including the localization of the different air quality markers on the map, the real-time air quality information for a particular location, the trend analysis of the different air quality parameters and the air quality forecast for a particular location. In this phase the different modules consisted in the mobile application will be tested. Figure 9 shows the sample output for real-time air quality information for a particular location.

Performance testing was carried out in order to determine the average time for data capture and transmission from the monitoring node to the base station, average time for data transmission from the base station to the cloud server and average time for data transmission from the cloud server to the mobile application. It should be noted that the time taken for data transmission is a function of the transmission medium (ZigBee, Wi-Fi, 3G). The average time for data capture is about 1000 ms between each retrieval, and the average time taken for data transmission from the monitoring node to the base station is 2000 ms. The average time for the AirVisual mobile application to download the air quality data from the cloud server for a particular location is less than 10,000 ms measured using an android monitor.

For the user acceptance test of the AirVisual mobile application, a group of 50 potential users were chosen to test the different features of the application. The users were briefly introduced to the AirVisual mobile application, and they were requested to use the application for a period of 2 weeks after which their feedback were collected. Table 2 shows a compilation of the general feedback received. While most of the users found the application easy to use, some users had difficulties in interpreting the graphs.

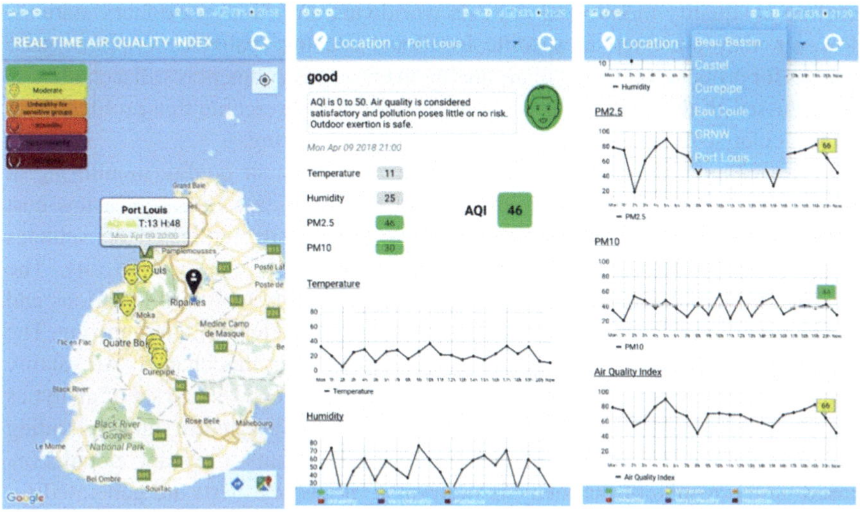

Fig. 9 Sample output for real-time air quality information for a particular location

Table 2 User feedback for AirVisual

Features	Feedback
Menus	Proper and simple buttons layouts. Easily interpretable options. Application logo was very well appreciated. Overall very positive feedback for the menu screen
Google Map AQI screen	Clear presentation of the AI for specific location. Some users had difficulties to interpret the AQI. The colour scheme used to represent air quality was well appreciated and helps to get a better understanding of the air quality. Some users felt the time to load air quality data from the cloud server was too long
Real-time AQI screen	The drop-down list eased navigation. The bar chart was preferred to the line graphs. Some users felt that there were too much information on a single screen
Trend analysis screen	The users found the trend analysis a very interesting and interesting feature. The bar chart were preferred to the line graphs. Some users had difficulties to interpret trends for the air quality
Notification screen	Many users were not able to use the notification feature, but they believed that it is very useful. It was suggested that proper instructions be included on how to set the notification

8 Conclusion

With the latest developments in sensor technologies, IoT systems have great potential to be used as effective solutions for air quality monitoring in cities on a real-time basis. Use of appropriate gaseous sensors with wireless communication capabilities can provide real-time ambient air quality information to the users.

Existing air quality monitoring systems worldwide are mostly stationary stations. However, they are fixed in specific locations, and the spatial coverage of such systems for proper monitoring of air quality in urban environment is still a challenge. As discussed in this chapter, one possible solution to overcome this problem is to use vehicular sensor networks for the air quality monitoring.

In this chapter, a vehicular IoT system for real-time air quality monitoring in smart cities is described. The main focus of this work was to design a low-cost IoT module for air quality monitoring that can be easily replicated on vehicles and provide good spatial coverage of the status of air quality in a city. The proposed system is an alternative to expensive air quality monitoring stations and provides fine-grained air quality data for better spatial and temporal monitoring. The vehicular-only deployment of the system is described in this chapter. Air quality sensor data are collected and transmitted to a remote server where a set of analytics tasks are performed in order to calculate the air quality index, thus providing real-time ambient air quality information to the local authorities and population. Based on the results of the analysis performed, necessary alerts are generated for abnormally high level of air pollution detected in a specific area. Moreover, a mobile application was developed for easy visualization of the air quality data. Through the mobile application, the general public is informed about the quality of the air in different regions of a city. The air quality data are presented visually using different coding schemes and format for easy interpretation and understanding. People may use the air quality mobile application to better plan their travel and take necessary precautionary measures.

References

1. Gubbi, J., Buyya, R., Marusic, S., & Palaniswami, M. (2013). Internet of Things (IoT): A vision, architectural elements, and future directions. *Future Generation Computer Systems, 29*(7), 1645–1660.
2. Dhingra, S., Madda, R. B., Gandomi, A. H., Patan, R., & Daneshmand, M. (2019). Internet of Things mobile-air pollution monitoring system (IoT-Mobair). *IEEE Internet of Things Journal, 6*(3), 5577–5584.
3. Balasubramaniyan, C., & Manivannan, D. (2016). Iot enabled air quality monitoring system (AQMS) using raspberry Pi. *Indian Journal of Science and Technology, 9*(39), 1–6.
4. Schneider, P., Castell, N., Vogt, M., Dauge, F. R., Lahoz, W. A., & Bartonova, A. (2017). Mapping urban air quality in near real-time using observations from low-cost sensors and model information. *Environment International, 106*, 234–247.
5. De Filippis, T., Rocchi, L., Rapisardi, E., Zaldei, A., Vagnoli, C., & Martelli, F. (2013). SensorWebBike: A participatory environmental urban monitoring. In *International Conference on Urban Climate and History of Meteorology* (pp. 243–243).
6. Vagnoli, C., Martelli, F., De Filippis, T., Di Lonardo, S., Gioli, B., Gualtieri, G., et al. (2014). The SensorWebBike for air quality monitoring in a smart city. In *IET Conference on Future Intelligent Cities*.
7. Arroyo, P., Herrero, J. L., Suárez, J. I., & Lozano, J. (2019). Wireless sensor network combined with cloud computing for air quality monitoring. *Sensors, 19*(3), 691.

8. Environment Statistics. (2014). Environment statistics year 2014. *Statistic Mauritius*. Republic of Mauritius Portal.
9. EPA. (2008). National environmental standards for air. *Environment Protection Act*. Republic of Mauritius Portal.
10. USEPA. (2006). Guideline for reporting of daily air quality: Air quality index (AQI). *United States Environmental Protection Agency*, EPA-454/B-06-001.
11. ThingSpeak. (2019). ThingSpeak IoT analytics platform service. *The MathWorks, Inc.* Available from: https://thingspeak.com/pages/learn_more.
12. Mahdavinejad, M. S., Rezvan, M., Barekatain, M., Adibi, P., Barnaghi, P., & Sheth, A. P. (2018). Machine learning for Internet of Things data analysis: A survey. *Digital Communications and Networks, 4*(3), 161–175.

Renewable Mobility in Smart Cities: The MOVESMART Approach

Damianos Gavalas, Kalliopi Giannakopoulou, Vlasios Kasapakis, Dionisis Kehagias, Charalampos Konstantopoulos, Spyros Kontogiannis, Damianos Kypriadis, Grammati Pantziou, Andreas Paraskevopoulos, and Christos Zaroliagis

1 Introduction

The provision of personalized mobility services in urban environments that combine transportation modalities (electromobility, public transportation, etc.) together with the advent of modern transport models like bike and car sharing is essential for creating efficient sustainable mobility chains for travelers. Such a development presents many challenges related to the route planning in urban traffic networks as well as the offering of innovative features in urban mobility.

D. Gavalas (✉)
Department of Product & Systems Design Engineering, University of the Aegean, Syros, Greece

Computer Technology Institute, Patras, Greece
e-mail: dgavalas@aegean.gr

K. Giannakopoulou · A. Paraskevopoulos · C. Zaroliagis
Computer Technology Institute, Patras, Greece

Department of Computer Engineering & Informatics, University of Patras, Patras, Greece
e-mail: gianakok@ceid.upatras.gr; paraskevop@ceid.upatras.gr; zaro@ceid.upatras.gr

V. Kasapakis
Computer Technology Institute, Patras, Greece

Department of Cultural Technology & Communication, University of the Aegean, Mytilene, Greece
e-mail: v.kasapakis@aegean.gr

D. Kehagias
Centre for Research and Technology Hellas/Information Technologies Institute (CERTH/ITI), Thessaloniki, Greece
e-mail: diok@iti.gr

M. M. Banat, S. Paiva (eds.), *Smart Technologies for Smart Cities*, EAI/Springer Innovations in Communication and Computing,
https://doi.org/10.1007/978-3-030-39986-3_7

Route planning in urban traffic networks poses two main requirements: (i) structure and maintenance of urban traffic information, so as to support the provision of time-dependent route-planning profiles in real time, overcoming the limited computational capabilities of navigation devices, and (b) intensive monitoring and assessment of the real-time status of the network via live traffic reports and capability of both predicting short-term disruptions and suggesting in real time traffic contingency plans to the users.

Modern urban mobility requirements call for comprehensive solutions which tackle the following challenges:

- *Renewable Mobility on Demand*: End users of eco-friendly mobility services should be offered intermodal mobility options, including electric vehicle (EV) transportation. EVs supplement public transit systems, offering flexibility for the first and last leg of a trip, which implies that an ideal solution would involve the deployment of urban car-sharing fleets in the vicinity of public transportation hubs.
- *Incentivized Vehicle Sharing (IVS)*: Incentivize travelers to pick up shared vehicles (mostly bicycles and electric cars) in low-demand places and drop them off in high-demand places, so as to maintain a balanced distribution of vehicles across the system's operating area, thus addressing future user demand. For this, a traveler-rewarding scheme is required, in order to induce well-behaving strategies adopted by the travelers. The design of such an incentive scheme for travelers is a major algorithmic challenge, since such a rewarding scheme must also be fair and credible and respect the travelers' privacy. Additionally, undesired behaviors should be excluded, for example, dropping off an EV at a docking station that possesses no recharging equipment.
- *Integrated Personal Mobility*: Promote and provide intermodal mobility options to travelers (with emphasis on eco-friendly means of transport, such as bicycles and EVs), which offer the option to incorporate various means of transportation in an efficient mobility chain. This is consistent with the next-generation bike-sharing systems deployed in major European cities.

C. Konstantopoulos · D. Kypriadis
Computer Technology Institute, Patras, Greece

Department of Informatics, University of Piraeus, Piraeus, Greece
e-mail: konstant@unipi.gr; dkypriad@unipi.gr

S. Kontogiannis
Computer Technology Institute, Patras, Greece

Department of Computer Science & Engineering, University of Ioannina, Ioannina, Greece
e-mail: kontog@cs.uoi.gr

G. Pantziou
Computer Technology Institute, Patras, Greece

Department of Informatics, University of West Attica, Athens, Greece
e-mail: pantziou@uniwa.gr

The EU-funded research project MOVESMART[1] addresses the aforementioned challenges and existing weaknesses of current systems for storing and maintaining traffic information, as well as in providing efficiently time-dependent route plans in large-scale urban traffic networks. This is achieved by adopting a holistic approach (and integrated platform) for the provision of renewable personal mobility services, leveraging crowdsourcing data, tools for collecting real-time information by multimodal travelers, and traffic prediction mechanisms (TPM). The goal is to guarantee real-time responses to renewable (on-demand) mobility queries for efficient multimodal route planning that are time-dependent as well as sensitive to aperiodic incidents (such as car accidents, sudden weather conditions) and traffic prediction forecasts.

The MOVESMART platform includes a backend system and front-end services. The *backend* comprises a hierarchical cloud-based urban traffic infrastructure. MOVESMART generates real-time urban traffic data and metadata through its live traffic logging system on the cloud, hosted by its urban traffic knowledge base (UTKB). One of the innovative aspects of MOVESMART is the provision of multimodal mobility services addressing the needs and limitations imposed by EVs in addition to other modes of public transport. An Energy Efficiency Assessment Module undertakes the estimation of the environmental footprint of the calculated routes, thus providing a means for the holistic assessment of the provided solutions from an energy-saving point of view, including EVs.

In addition, the MOVESMART delivers end-user (*front-end*) services with innovative features:

(i) *Renewable Mobility on Demand*, by providing the travelers with all necessary information required so that they can create an eco-friendly mobility chain consisting of EVs and public transportation media and by allowing them to decide "on the spot" which transportation mode they wishes to choose for the next leg of their trip.
(ii) *Incentivized Vehicle Sharing* through the development of innovative algorithms that ensure EV booking, balanced distribution of EVs across stations, rewarding schemes that enforce social behavior, and recommendations for appropriate pickup/drop-off charging stations.
(iii) *Integrated Personal Mobility* by providing the necessary accounting information to an integrated mobile application which enables "door-to-door" dynamic route planning and navigation, based on real-time data and accurate predictions.

This chapter reports on the main aspects and contributions of the MOVESMART ecosystem, including its key architectural entities, main innovative features, and end-user services. It is noted that MOVESMART services have been validated during the official project pilots in Vitoria-Gasteiz (Basque Country, Spain) and Pula-Pola (Croatia). Among other renewable mobility options, the two pilot sites

[1] MOVESMART project page: http://movesmartfp7.iti.gr/

allowed the validation of the end-user services on a free-floating e-scooter-sharing system and a station-based e-bike-sharing system, respectively.

The remainder of the chapter is structured as follows: Section 2 provides a brief overview of the MOVESMART approach. Section 3 describes the key aspects and novel features of the MOVESMART backend services. Section 4 discusses the main components of the MOVESMART end-user renewable mobility services. Section 5 presents the integrated MOVESMART mobile application. Finally, Section 6 concludes the chapter.

2 Overview of the MOVESMART Approach

Due to the computationally demanding procedures for harvesting, processing, and maintaining information, a key decision in MOVESMART has been to create an urban traffic knowledge base (UTKB). UTKB is hosted in a cloud of urban traffic servers and manages all the preprocessed data updates. The token is then passed to mobile devices (either built-in car devices or smartphones), which dispatch route-planning profile queries to be handled locally according to the individual needs of the end user and the options s/he is provided.

MOVESMART also provides a synchronization mechanism in order to deal with potential disconnections of the mobile devices; the mechanism regularly updates the local devices with time-dependent snapshots provided by the UTKB. To avoid the aforementioned weaknesses on traffic monitoring and digestion, MOVESMART employs a novel crowdsourcing scheme to involve travelers in traffic information collection, making use of ad hoc traffic-sensing entities. These entities may be either in-car or mobile devices (PNDs or smartphone applications) which enable either periodic or instantaneous on-route traffic reports, on behalf of the end users. The objective is to cover the entire metropolitan network, or at least its most heavily used part, so that is complementary to that of the fixed traffic-sensing devices. The ad hoc traffic-sensing devices may issue either periodic reports (e.g., average-speed reports) on specific road segments or some warnings for emergent incidents such as a car accident which temporarily blocks some particular road segments, or a sudden climate event that results to an unforeseen congestion. The addressing of these unexpected events requires the consideration of dynamic networks, in which particular elements may suddenly become temporarily unavailable for a certain period of time. Finally, each end user may also provide feedback through submitting post-route satisfaction reports on the quality of the route plans proposed by the system, in a like/dislike fashion.

The rationale of the entire MOVESMART system is illustrated in Fig. 1: There is an everlasting life cycle of traffic information *collection* via the fixed and the ad hoc traffic-sensing entities, *prediction* of emergent (aperiodic) incidents and provision of appropriate alerts via the traffic prediction mechanism (TPM), *adaptation* of the maintained data structures in the UTKB, *computation* of time-dependent route

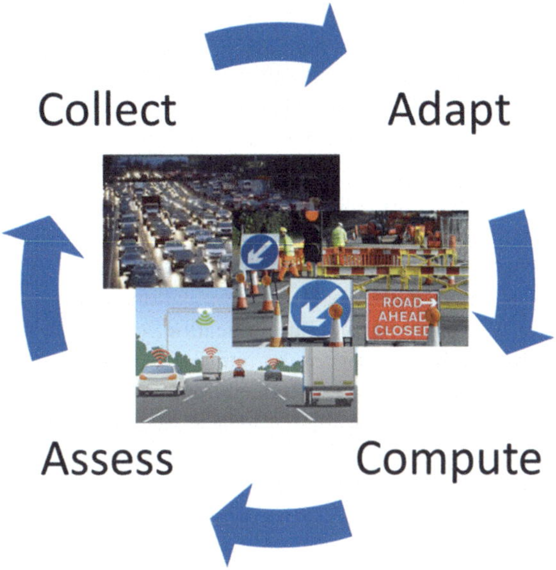

Fig. 1 The MOVESMART traffic information life cycle

plans and incident-dependent route replans, and finally post-trip *assessment* of the suggested route (re-)plans directly by the end users.

3 Backend Services of the MOVESMART Platform

3.1 Crowdsourcing

The *crowdsourcing platform* provides an open crowdsourcing framework for collecting and managing traffic data. The collection of traffic data involves (a) on-route data, related with user's location and speed in real time; (b) post-route data, concerning the feedback information that users send in order to evaluate nearby reports and ensure collected data quality; and (c) emergency data related with unpredictable events like car accidents, works, or extreme weather phenomena that directly affect traffic conditions.

The idea of hiring and tracking users with GPS-enabled devices has been employed to obtain live traffic data and improve the accuracy and reliability of routes recommended by car navigation systems. This is especially useful to yield traffic information on road network areas not covered by fixed traffic-sensing infrastructure. MOVESMART's specific contributions include:

- Enabling a mechanism for collecting real-time emergency reports of disruptions, transmitted to end users (based on the significance of the reports and the reliability of the reporters). In this respect, this mechanism involves a particular module

capable of estimating the reliability of collected data as a means to ensuring the quality and reducing the impact of malicious users. The integration of the crowdsourcing functionalities with the report evaluation algorithm has been a critical challenge but also the main success factor of the whole work. This is highly important for the assessment and future improvement of a crowdsourcing system.
- Using the post-route traffic information to update the traffic prediction module and thus assessing its quality. This is highly important for the assessment and future improvement of a traffic prediction system.
- Development, release, and evaluation of the crowdsourcing service, initially as a standalone mobile application, called "Live Traffic Reporter" (LTR) and eventually as a part of the integrated "MOVESMART"[2] application. Alongside this endeavor, several challenges have been tackled to achieve the optimal result. Cutting-edge technologies have been used in all stages and levels of development. Our aim has been to use the most fitting and effective technologies as a means to ensuring that the developed application would meet all modern standards for mobile applications.

As stated above, a major contribution of the crowdsourcing mechanism implemented as part of the MOVESMART project is the development of a prototype reliability assessment mechanism used to validate the crowdsourced information [1]. Based on the fact that the crowdsourcing mechanism collects data from various users who may send false, inaccurate, or misleading information, there is a need for evaluating all collected data so as to ensure their quality. To this end, we have developed a user credibility algorithm which aims to evaluate the trustworthiness of each user as well as an algorithm that evaluates the reliability of each submitted report (emergency data). The innovative features that have been implemented include the use of the user's feedback on previous users' reports for the calculation of the credibility, as well as a new formula to calculate the overall score that is used for detecting "spam" reports.

Figure 2 illustrates a high-level architectural diagram of the crowdsourcing service. All in all, the crowdsourcing system consists of a user credibility management system that includes also a feedback database and a validation component used to validate crowdsourced information based on user credibility. The crowdsourcing module uses the crowdsourced information, according to the outcome of the data evaluation mechanism.

3.2 *Traffic Prediction*

Vehicular traffic prediction at different spatiotemporal instances is considered as one of the primary means for enhancing the performance of decision support

[2]https://play.google.com/store/apps/details?id=gr.iti.Movesmart&hl=el

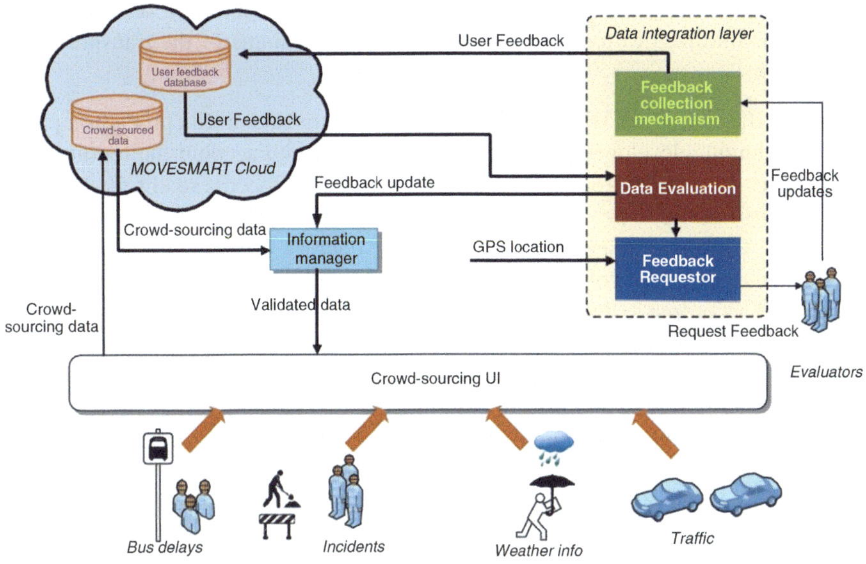

Fig. 2 A high-level architectural diagram of the crowdsourcing service

systems in modern intelligent transportation system (ITS). The traffic prediction service is incorporated in the *Renewable Mobility on Demand* asset, a realization of Renewable Mobility on Demand based on the corresponding MOVESMART service. This service is also included in the *Transport Data Repository* (accessible on the cloud), a set of cloud-provided services in the domain of transport which will provide traffic prediction services based on historical and crowdsourced data.

The MOVESMART's specific contributions in the traffic prediction service include the following:

- Developing novel algorithmic techniques to incorporate users' semantic feedback about traffic conditions and abnormal phenomena (e.g., due to weather alterations and accidents) into the traffic prediction scheme. The most effective existing traffic prediction methodologies have been enhanced so as to adapt upon the notification of traffic incidents. Moreover, suitable interfaces have been developed for efficiently mapping all the abnormal traffic phenomena onto measurable and exploitable information; those interfaces cater for the convenient use of the tool (and hence the maximization of the users' participation) and the effective transformation of the users' feedback.
- Enabling monitoring and assessment of the real-time status of the network via live traffic reports and capability of both predicting short-term disruptions and suggesting in real-time traffic contingency plans to the users causing the minimum overall disturbance [2].
- Developing algorithmic techniques to smooth human feedback and disregard overly biased user reports, thereby enhancing the reliability of the traffic

prediction input. Third-party sources (e.g., weather forecast web services) have been also utilized to formulate a clear view on the status of the transportation network.

Graph-Based Lag-STARIMA (GBLS) [3], the innovative traffic prediction parametric technique developed in MOVESMART, takes into account the topology of the network for creating regions of roads correlated to a specific road for which we wish to perform traffic prediction. Another newly developed approach comprises a parametric technique based on clustering traffic data for creating various homogeneous road occupancy profiles. Data clustering is applied on different time periods (segments) according to the type of traffic (free flow vs. high congestion). These two approaches have been evaluated by means of a benchmarking environment, against a set of well-known techniques from the relevant literature. The dataset used for benchmarking purposes and also for training the two techniques has been derived from a network of inductive loop detectors in the pilot city of Vitoria-Gasteiz, Spain.

Figure 3 illustrates the total prediction accuracy results of the GBLS model, compared to the baseline solutions: k-nearest neighbor (kNN), random forest (RF), STARIMA, and clustering-based algorithms. The prediction accuracy metric used for evaluating the performance of the algorithms is the Root Mean Square Error (RMSE) due to its robustness to close-to-zero values. Because the data from Vitoria-Gasteiz concern traffic flow values, RMSE is measured in vehicles per minute.

An average error of less than one vehicle per minute for GBLS in weekdays is considered insignificant for a traffic network when the average flow is between 50 and 60 vehicles per minute. The same applies for weekends when the average flow is between 35 and 40 vehicles per minute.

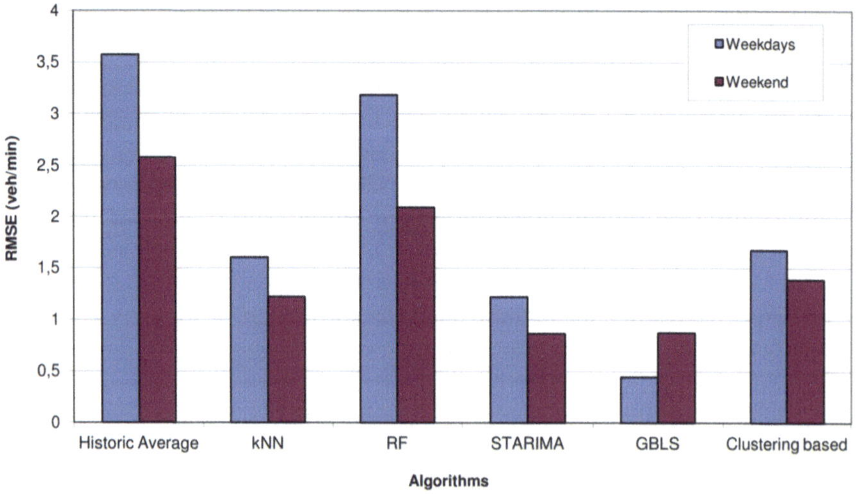

Fig. 3 Prediction accuracy results by the MAPE metric

3.3 UTKB and Time-Dependent Route Planning

The computation of time-dependent route-planning profiles in large-scale urban traffic networks represents a particularly challenging algorithmic task. MOVESMART's specific contributions include the following:

- Developing techniques for structuring and maintaining urban traffic information, so as to support the real-time provision of time-dependent route-planning profiles, and establishing a trade-off between required preprocessed space and query time.
- Developing new methods for graph preprocessing approaches that enable fast responses to shortest-path queries (i.e., distance oracles) in time-dependent road networks [4], by exploiting the inherent hierarchical structure of the MOVESMART system, overcoming the shortcomings of existing approaches, and thus contributing to the establishment of new research paradigms.

In particular, the urban traffic knowledge base (UTKB) acts as the pivotal point of all the traffic-related information (raw traffic data) aggregated by the MOVESMART system from external sources (e.g., from loop detectors, crowd of travelers), or service-oriented traffic metadata (e.g., travel-time functions between distant points, traffic-sensitive energy consumption factors) created and continuously updated within it. All these raw traffic data and metadata are then at the disposal of the core routing services of MOVESMART (i.e., the core time-dependent routing (TDR) service, the multimodal routing (MMR) service, and the electric vehicle routing (EVR) service), which run on a dedicated virtual machine (the UTKB server) of the MOVESMART's cloud. These services are in turn exploited by the more sophisticated personalized mobility and mobility-on-demand services.

A stream of temporal traffic–related information, which is produced by the crowdsourcing application Live Traffic Reporter (LTR), is also digested by the UTKB. This stream of temporal data provides real-time information about the actual status of the network and is aggregated at the UTKB as either traveler-initiated emergency reports (e.g., blockages of road segments due to car accidents) or automatically generated traffic prediction alerts (because of prediction of some significant deviation from the nominal behavior of a road segment). Appropriate temporal metadata are also maintained within the UTKB in real time and are interpolated with the abovementioned raw traffic data and metadata, in order to support the sophisticated time-dependent, live traffic–aware routing services of the MOVESMART system.

Finally, periodically updated snapshots of the traffic-related data are disseminated to the travelers' smartphone devices, in support of elementary routing services, as a contingency plan in case of connectivity loss with the MOVESMART system. Figure 4 provides an architectural overview of the UTKB functionalities.

The Time-Dependent Route Planning service is included in the *Renewable Mobility on Demand* asset, a realization of Renewable Mobility on Demand based

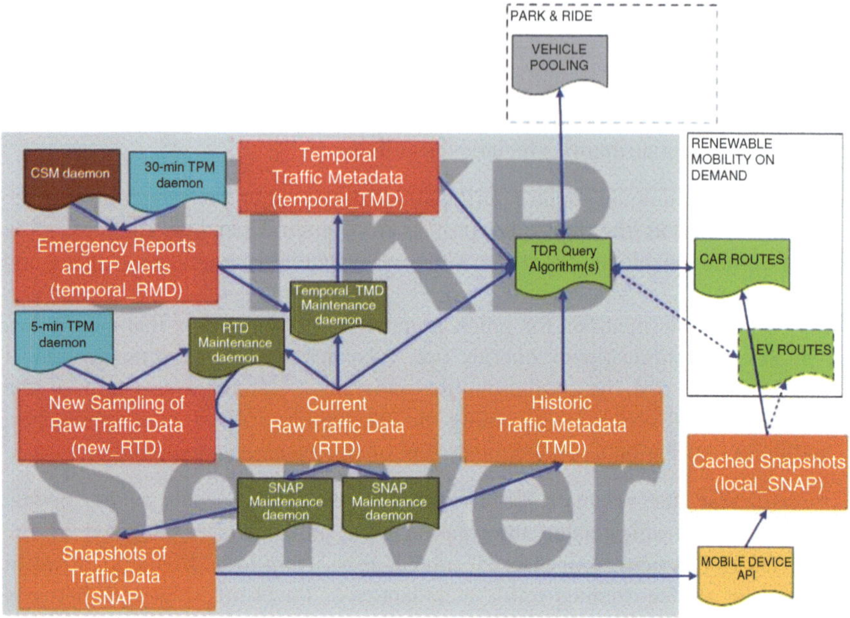

Fig. 4 Overall architecture of UTKB-related services

on the corresponding MOVESMART service. This service is also included in the *Transport Data Repository* (accessible on the cloud), a set of cloud-provided services in the domain of transport. Finally, Time-Dependent Route Planning is exposed through the *Innovative Route Planner*, a generic platform for renewable mobility services based on novel time-dependent routing approaches which take into account live traffic reporting.

The aforementioned Time-Dependent Route Planning services have already been incorporated in the integrated MOVESMART application (available from Google Play).

3.4 Energy Efficiency Assessment Module

Energy efficiency assessment methodologies aim to provide novel algorithmic approaches and services to support the energy-efficient intermodal mobility of individuals (with emphasis on environmentally friendly means of transportation, such as bicycles and electric transportation media), in order to assist travelers in combining various means of transportation in an energy-efficient way. However, the design of such a service for travelers represents a major algorithmic and technical challenge, since various aspects, models, and parameters of energy efficiency need to be considered within the renewable mobility scenario in order to achieve the requested energy-efficient multimodal route planning. There are certain different ways/methodologies and corresponding simulation tools (i.e., COPERT, ADVISOR, GREET, etc.) to calculate the emissions and the energy efficiency of conventional vehicles as well as of EVs (along with that of the electromobility, in general) while in action or when parked. MOVESMART's specific contributions include:

- Developing novel algorithmic approaches to compute emissions and energy consumption of journeys that combine different modes of transport, including conventional and electric vehicles, buses, and trams.
- Providing the required expertise for the selection/development/combination of the proper methodologies in order to consider the characteristics of the applied EVs in the MOVESMART project for the energy efficiency as well as the calculation of the emissions.

In the field of energy efficiency assessment, MOVESMART proposed methods to assess and compare the environmental impact of multimodal journeys, in terms of both energy consumption and emissions. On the one hand, the methods take energy operational processes into account, that is, producing, transporting, manufacturing, and distributing fuels and electricity. On the other hand, the methods consider vehicle operational processes, evaluating the efficiency of the vehicles themselves.

Figure 5 shows the comparative results of energy consumption when considering several driving cycles,[3] after simulating the performance of the test vehicle with the

Fig. 5 Validation results of energy consumption model for electric cars

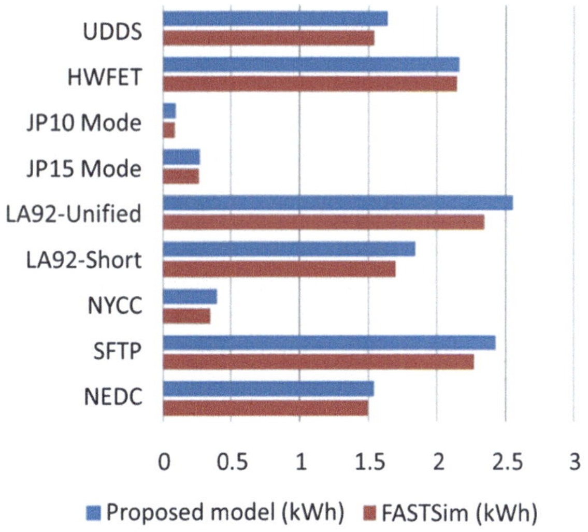

[3]A driving cycle is a series of data points representing the speed of a vehicle versus time. Driving cycles are produced by different countries and organizations to assess the performance of vehicles in various ways, for example, fuel consumption and polluting emissions.

proposed model and FASTSim [5]. A careful examination of the results reveals that the maximum difference (in absolute values) between the two approaches is observed in the LA92 Unified cycle, which is the longest one in terms of time and represents aggressive city-driving conditions, while minor differences are observed in the case of smoother driving cycles, such as JP 10 Mode, JP 15 Mode, and HWFET. In general, the results obtained indicate that the output of the developed model follows with sufficient accuracy the FASTSim output under different (simulated) driving conditions.

The MOVESMART Energy Efficiency Assessment service is already incorporated in the integrated MOVESMART application and is also available in the *Transport Data Repository* (accessible on the cloud), a set of cloud-provided services in the domain of transport.

4 End-User Renewable Mobility Services

The combination of urban transportation modalities like electromobility, bicycles, and public transportation along with the emergence of transportation models like bicycle and car sharing is considered as effective means to create eco-friendly mobility chains in urban transfers.

MOVESMART has delivered an integrated personalized mobility service to provide multimodal routes that combine various means of transportation (including EVs, public transportation, cycling, and walking) [6, 7]. Among others, the service addresses the scenario of EV drivers/commuters receiving recommendations for available drop-off charging stations, suitably located nearby public transit hubs, ensuring fastest end-to-end time-dependent travel times.

The derived algorithmic solutions already support an incentivized vehicle-sharing service, wherein users are motivated to pick-up vehicles (bicycles or EVs) from low-demand zones and drop them off in high-demand areas [8–10]. For this purpose, techniques inspired from game-theoretic and operational research approaches have been implemented. Several models and algorithms enabling renewable personal mobility in urban networks have been developed, described, and evaluated. The outcome clearly broadens the state-of-the-art in our three main topics: (i) eco-friendly multimodal route planning, (ii) vehicle-sharing systems with customer incentivization, and (iii) energy efficiency assessment of multimodal journeys.

4.1 Eco-Friendly Multimodal Route Planning

In the field of route planning, MOVESMART introduced time-dependent speedup techniques [11] that are robust against user-dependent metric changes and enable fast near-optimal queries, with quick integration of user preferences, live traffic, and traffic predictions [12].

Moreover, an approach for quickly computing realistic pedestrian routes has been studied [13]. Building upon this study, geometric algorithms to automatically augment the street network with sensible sidewalks and edges in plazas have been developed, making it possible to walk across them in a natural way. Figure 6 shows an example in the city of Karlsruhe, Germany. It highlights the importance of, both, the presence of sidewalks and being able to route across walkable areas. Compared to other popular routing services, MOVESMART routing has no unnecessary street crossings (because of generated sidewalk data), and the plaza is traversed in a natural way.

Regarding EV mobility, MOVESMART dealt with computing sets of routes for electric vehicles that trade travel time and energy consumption via Pareto optimization [14]. To the best of our knowledge, in the process, we have been the first to explicitly consider driving road segments at different speeds (below the limit) in order to save energy. Moreover, MOVESMART introduced a new approach to route planning for EVs that computes shortest feasible paths with charging stops, minimizing overall trip time [15]. To alleviate range anxiety, displaying the remaining cruising range of an electric vehicle has been proposed, along with a portfolio of speedup techniques for the resulting isochrone problem. Finally, the project extended previous work to provide an integrated multimodal routing solution for the pilot applications, encompassing EV mobility (shared scooters and bikes), walking, and public transportation.

Figure 7 demonstrates the Multimodal Route Planner (MMR) in action. Multimodal Route Planner answers shortest-path queries, taking multiple modes of transportation into account. It always considers public transit and (unrestricted) walking, as well as station-based rental bicycles and free-floating rental scooters, depending on the city. To our knowledge, the most popular map services like Google Maps and OSM do not support so far park-and-ride scenarios to pair transit directions with biking and vehicle-sharing options.

An extensive pilot of MMR has been carried out in the cities of Vitoria-Gasteiz (Spain), which supported e-scooters, and in Pula-Pola (Croatia) which supported e-bicycles.

Further enhancements of the aforementioned renewable mobility services have been implemented through the algorithmic developments in [16–18].

4.2 *Incentivized Vehicle Sharing*

The major operational challenge dealt with in vehicle-sharing systems is the so-called asymmetric demand-offer problem which occurs when a large percentage of vehicles is concentrated to certain stations/areas of lower mobility demand (cold spots) while high demand stations/areas (hot spots) lack sufficient vehicle stock to satisfy user requests [19]. To this end, several vehicle-sharing systems opt to properly incentivizing travelers to pick-up shared electric vehicles, such as bicycles, in low-demand places and drop them off in high-demand places. For this,

(a)

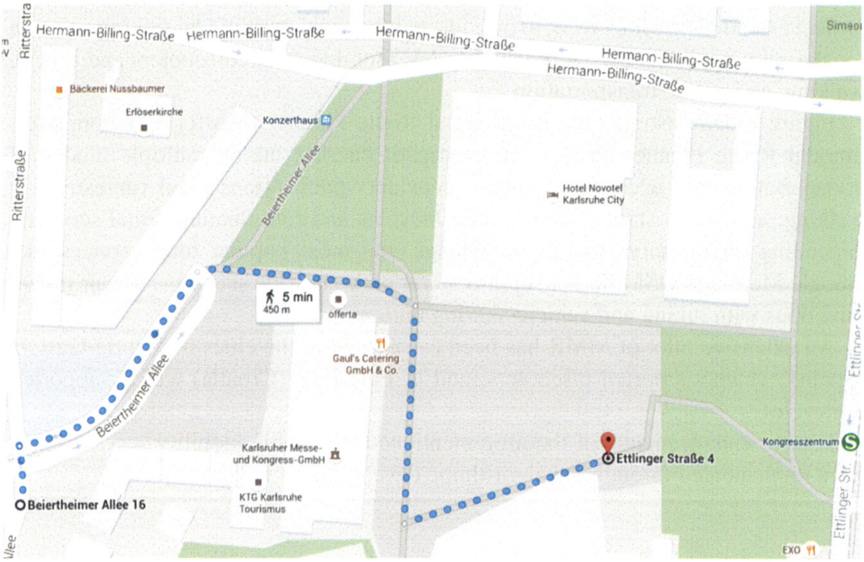

(b)

Fig. 6 Computation of pedestrian routes: (**a**) OpenRouteService; (**b**) Google Maps; (**c**) Nokia's HERE; (**d**) the MOVESMART approach

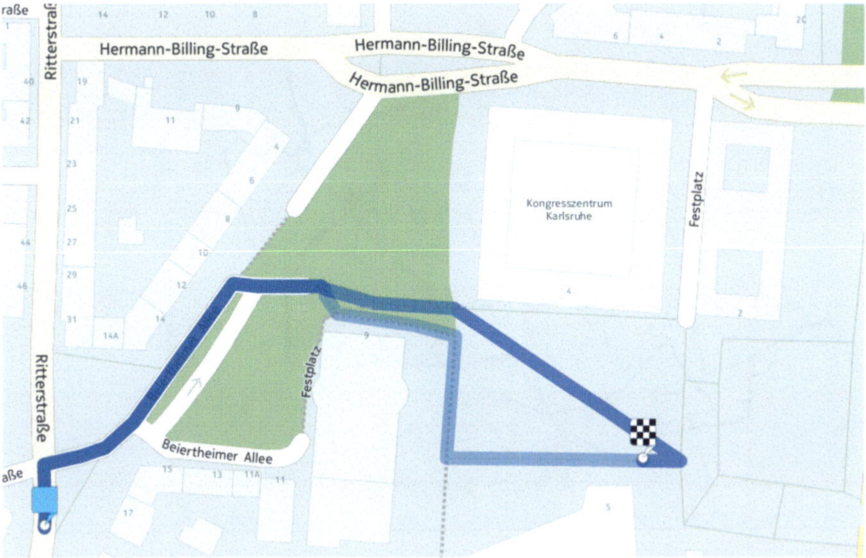

(c)

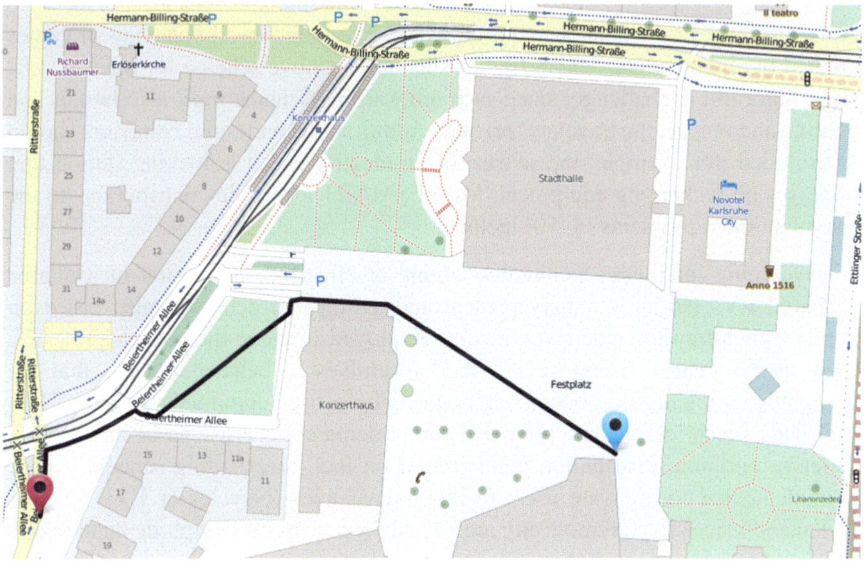

(d)

Fig. 6 (continued)

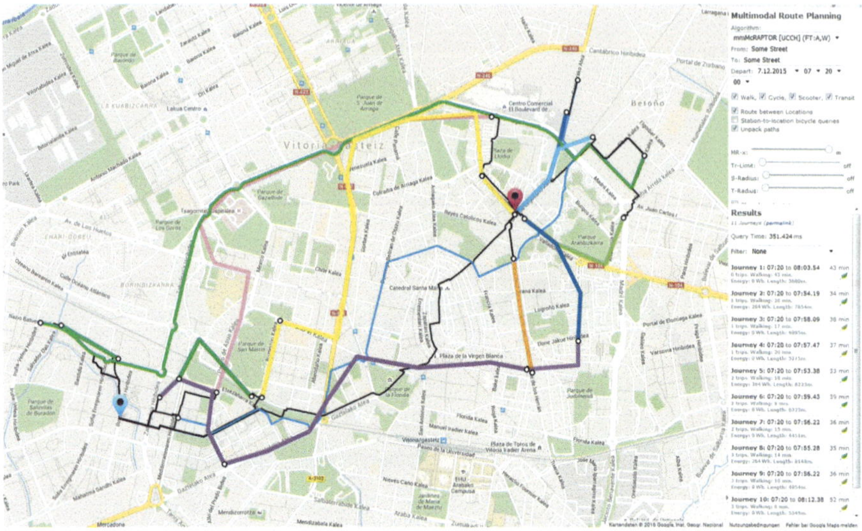

Fig. 7 Multimodal Route Planner (MMR) example from the pilots in the city of Vitoria-Gasteiz

a traveler-rewarding scheme is required, in order to induce well-behaving strategies adopted by the travelers. The design of such an incentive scheme for travelers is a major algorithmic challenge, since such a rewarding scheme must also be fair and credible and respect the travelers' privacy. Additionally, undesired behaviors should be excluded, for example, the release of an electric car at a docking station that possesses no recharging equipment. MOVESMART's specific contributions to the aforementioned problems [8–10] include:

- Developing intelligent techniques aiming at effective distribution of available shared EVs (so as to satisfy present and anticipated demand), also based on punishment/reward schemes typically employed in game-theoretic approaches. In such schemes, users are provided incentives to behave in ways that are socially efficient (i.e., beneficial to the whole user community). In the EV-sharing scenario, past and predicted future demand patterns can be analyzed to derive optimum distribution plans; based on that, a rewarding scheme can be implemented to motivate users to pick-up vehicles (bicycles or EVs) in low-demand places (e.g., remote city areas) and drop them off in high-demand places (e.g., nearby transport hubs).
- Developing algorithms that derive recommendations for appropriate pickup/drop-off charging stations (based on the available vehicle's state of charge and the user's willingness to take advantage of the rewarding scheme). A side objective of this task is to measure the long-term efficiency of EV sharing networks, namely, to what extent customer demand is satisfied and how well shared EV networks integrate with public transit networks. Both aspects are largely determined by trends in spatial–temporal distribution of demand and the distribution of parking/charging facilities in the urban space.

- Investigating novel algorithmic approaches that make use of historical demand and traveler mobility patterns in order to optimize the deployment of parking/charging facilities. Variations of the Maximal Expected Coverage Relocation Problem (the aim is to provide relocation strategies for public service facilities covering the overall demand) have been investigated to meet this objective.

In the field of vehicle sharing, MOVESMART developed new algorithmic incentivization schemes, enabling cost-effective customer-based redistribution of shared vehicles [9, 10]. The schemes offer a price reduction or credit to customers for delivering vehicles at empty stations. These schemes have been extensively evaluated using discrete-event simulations. Moreover, mathematical models have been developed for the strategic planning of vehicle-sharing systems [8]. These models determine the optimal number and location of stations, as well as the optimal fleet size and distribution of shared vehicles to stations. Two case studies were conducted using the new models: We planned a bicycle-sharing system in the city center of Athens and a car sharing system in the metropolitan area of Athens.

Figure 8 illustrates different options for a user reservation request. Suppose that a user is willing to participate in the incentivization scheme, who wishes to make a trip from a starting point A to a destination B. Two different options may be offered to the user: the first from station S_1 to station S_3 and the second from station S_1 to station S_2, the former being the shortest in time and, therefore, more convenient to the user. Assuming that S_2's occupancy is lower than S_3's occupancy, the second option is preferable to the Incentivized Vehicle Sharing (IVS) system. Therefore, the IVS may offer incentives to the user to compensate for the longer trip. If the user accepts the offered reward (e.g., price reduction for her trip), she will contribute in a more balanced distribution of vehicles across the system's operational area; therefore, she will support the capacity of the system to satisfy future user demand and will reduce the operational costs associated with vehicle relocations typically undertaken by employees of the operator [20].

The MOVESMART *Incentivized Vehicle Sharing service* provides a map-based interface that offers an incentivizing scheme to handle the users' reservation requests

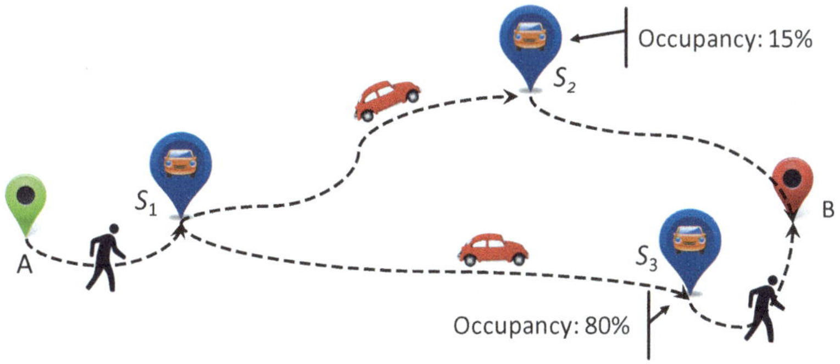

Fig. 8 Different options for a user request

for free-floating rental electric scooters (tested in the city of Vitoria-Gasteiz), or station-based rental electric bikes (tested in the city of Pula-Pola).

5 The Integrated MOVESMART Mobile Application

Building upon the aforementioned algorithmic and technical developments, MOVESMART has implemented an integrated mobile route planner application that seeks to address the problem of providing time-dependent route-planning profiles in large-scale urban traffic networks, promoting the eco-friendliest alternatives. Innovative functionalities like the Renewable Multimodal Mobility Route Planner, the crowdsourcing-based traffic prediction service [21], or the Incentivized EV Vehicle Sharing service have been incorporated into the application.

MOVESMART's specific contributions [6, 7] include:

- Development of an innovative mobile route planner application that seeks to provide the most efficient route from one point to another with respect to energy consumption and, therefore, leads to decreased emission of pollutants. For this purpose, the application integrates the crowdsourcing-based traffic prediction along with the energy-efficient Multimodal Route Planner [21]. By considering different modes of transport, the MOVESMART application provides transport route alternatives ranked not only by the estimated journey time but also by their environmental impact, prioritizing those which are more environmentally friendly, such as public transport or electric vehicles. Thus, it represents a tool which aligns with the climate protection strategies and improves energy efficiency in the mobility sector.
- Employment of persuasive methods to motivate behavioral change toward the adoption of eco-friendlier travel habits. When designing a product, or an interface, the visual presentation strategies influence user decisions. The user interface of the MOVESMART route planner client adopts mobile application usability principles to facilitate users on submitting trip requests and accurately perceiving the provided output. Additionally, the information visualization design employs persuasive technology strategies [22] to motivate users to prefer eco-friendlier options over others.

Aspects like portability, user privacy, and usability have been taken into account throughout the design and development of the integrated application. A key aim of MOVESMART has been to utilize the most appropriate and effective technologies as a means to ensure that the developed application would meet all modern standards for mobile application development.

The MOVESMART application is available at Google Play Store.[4] The mobile application as well as its corresponding functionalities (front-end and backend)

[4]https://play.google.com/store/apps/details?id=gr.iti.Movesmart&hl=el

have also been optimized and parameterized based on preliminary feedback and comments by real users.

Figure 9 illustrates the crowdsourcing service as a part of the integrated MOVESMART mobile application. Figures 9a, b present the procedure for submitting a traffic report, while Fig. 9c shows how a user is invited to validate reports submitted by others.

Figure 10 illustrates the Renewable Multimodal Mobility service as a part of the integrated MOVESMART mobile application.

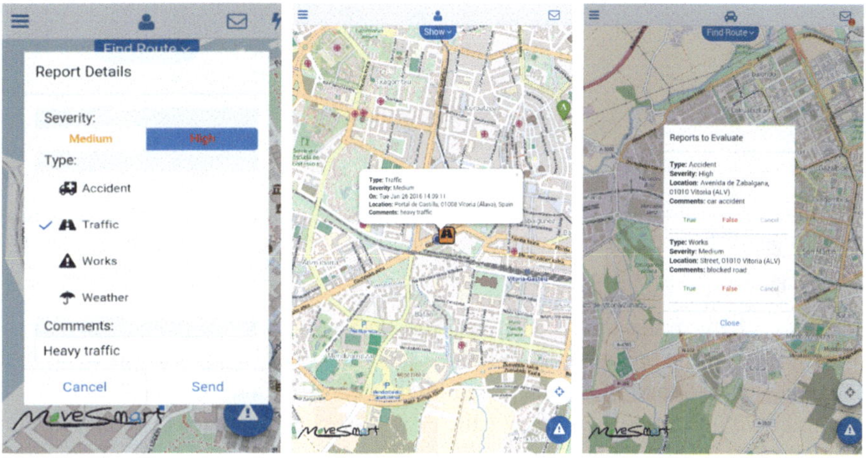

Fig. 9 Crowdsourcing service as a part of the integrated MOVESMART mobile application

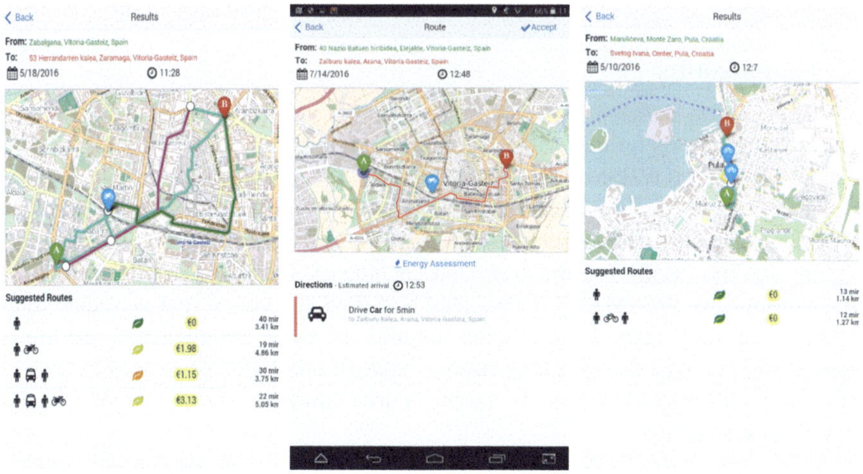

Fig. 10 Renewable Multimodal Mobility service as a part of the integrated MOVESMART mobile application

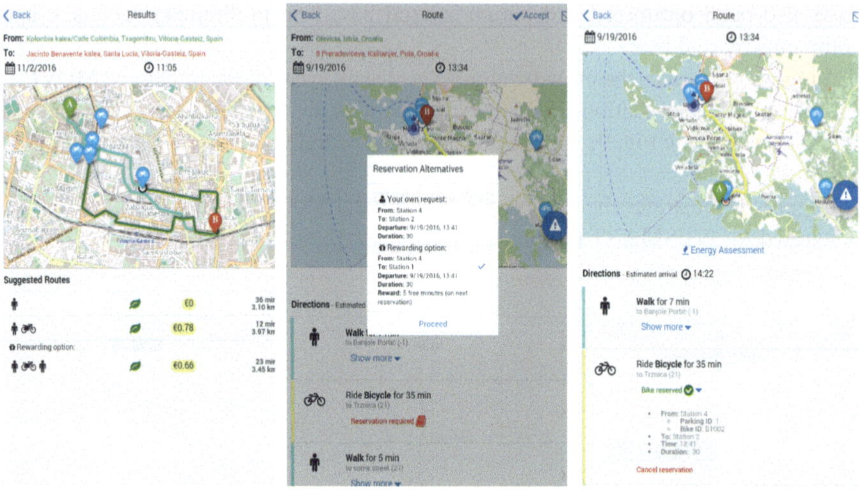

Fig. 11 Incentivized Vehicle Sharing service as a part of the integrated MOVESMART mobile application

Figure 11 illustrates the Incentivized Vehicle Sharing service (where the user is offered a rewarding option to drop off the e-scooter on a location different to the one she indicated) as a part of the integrated MOVESMART mobile application.

Figure 12 illustrates the Energy Efficiency Assessment service (as a part of the integrated MOVESMART mobile application) providing the user with information about the energy consumption of a particular transport option.

6 Conclusions

This chapter reports on the main outcomes and research contributions of the EU-funded project MOVESMART. MOVESMART proposes a holistic approach as regards renewable mobility services in smart cities. The MOVESMART approach provides personal mobility services, leverages crowdsourcing techniques for collecting real-time traffic information, and encompasses effective traffic prediction mechanisms. MOVESMART guarantees real-time responses to renewable (on-demand) mobility queries for efficient multimodal route planning that are time-dependent as well as sensitive to aperiodic incidents and traffic prediction forecasts. The core of MOVESMART is a hierarchical urban traffic infrastructure maintained in a cloud architecture.

A key aspect of the MOVESMART infrastructure is the design and implementation of novel renewable mobility services which incorporate eco-friendly modes of transport, most notably (shared) electric vehicles. The project delivers a set of innovative cloud-based mobility services, such as Renewable Mobility

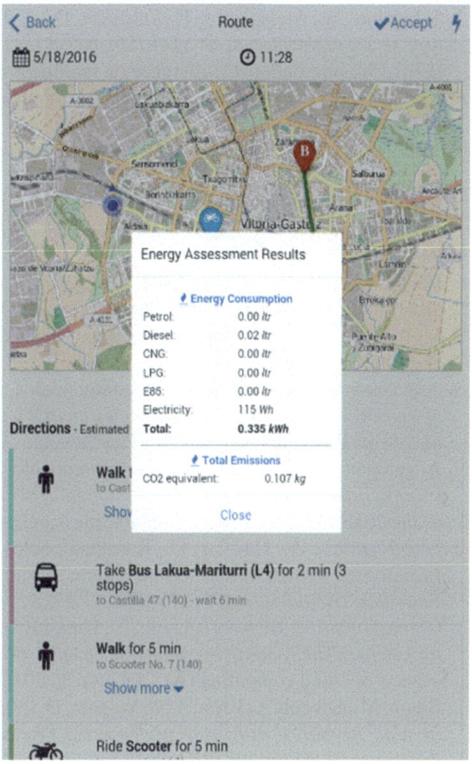

Fig. 12 Energy Efficiency Assessment service as a part of the integrated MOVESMART mobile application

on Demand, Incentivized Vehicle Sharing, and Integrated Personal Mobility, on which new business models are expected to be deployed involving service providers, EV manufacturers, and public authorities. The MOVESMART ecosystem has been validated in two different European cities with diverse transportation characteristics through extensive pilot tests.

Acknowledgments This work has been supported by the EU FP7 project MOVESMART (Grant Agreement No. 609026).

References

1. Salamanis, A., Drosou, A., Michalopoulos, D., Kehagias, D., & Tzovaras, D. (2016). A probabilistic framework for the reliability assessment of crowd sourcing urban traffic reports. *Transportation Research Procedia, 14*(2016), 4552–4561.
2. Salamanis, A., Kehagias, D. D., Filelis-Papadopoulos, C. K., Tzovaras, D., & Gravvanis, G. A. (2015). Managing spatial graph dependencies in large volumes of traffic data for travel-time prediction. *IEEE Transactions on Intelligent Transportation Systems, 17*(6), 1678–1687.

3. Salamanis, A., Meladianos, P., Kehagias, D., & Tzovaras, D. (2015). *Evaluating the effect of time series segmentation on STARIMA-based traffic prediction model*. Proceedings of the 2015 IEEE 18th international conference on intelligent transportation systems, pp. 2225–2230.
4. Kontogiannis, S., Papastavrou, G., Paraskevopoulos, A., Wagner, D., & Zaroliagis, C. (2017). *Improved oracles for time-dependent road networks*. Proceedings of the 17th workshop on algorithmic approaches for transportation modeling, optimization, and systems (ATMOS 2017): 4:1–4:17.
5. Brooker, A., Gonder, J., Wang, L., Wood, E., Lopp, S., & Ramroth, L. (2015). *FASTSim: A model to estimate vehicle efficiency, cost and performance (No. 2015-01-0973)*. SAE Technical Paper.
6. MOVESMART Deliverable D4.2.2: Renewable energy-efficient multimodal route planning algorithmic approaches. MOVESMART Project, July 2016.
7. MOVESMART Deliverable D4.3.2: Renewable personal mobility trip planner. MOVESMART Project, July 2016.
8. Angelopoulos, A., Gavalas, D., Konstantopoulos, C., Kypriadis, D., & Pantziou, G. (2016). *An optimization model for the strategic design of a bicycle sharing system: A case study in the city of Athens*. Proceedings of the 20th Pan-Hellenic conference on informatics (PCI 2016).
9. Angelopoulos, A., Gavalas, D., Konstantopoulos, C., Kypriadis, D., & Pantziou, G. (2016). *Incentivization schemes for vehicle allocation in one-way vehicle sharing systems*. Proceedings of the 2016 IEEE international smart cities conference (ISC2 2016).
10. Angelopoulos, A., Gavalas, D., Konstantopoulos, C., Kypriadis, D., & Pantziou, G. (2018). Incentivized vehicle relocation in vehicle sharing systems. *Transportation Research Part C: Emerging Technologies, 97*, 175–193.
11. Giannakopoulou, K., Kontogiannis, S., Papastavrou, G., & Zaroliagis, C. (2016). *A cloud-based time-dependent routing service*. Proceedings of the international workshop of algorithmic aspects of cloud computing (ALGOCLOUD 2016), pp. 41–64.
12. Baum, M., Dibbelt, J., Pajor, T., & Wagner, D. (2016). *Dynamic time-dependent route planning in road networks with user preferences*. Proceedings of the international symposium on experimental algorithms (SEA 2016), pp. 33–49.
13. Andreev, S., Dibbelt, J., Nöllenburg, M., Pajor, T., & Wagner, D. (2015). *Towards realistic pedestrian route planning*. Proceedings of the 15th workshop on algorithmic approaches for transportation modelling, optimization, and systems (ATMOS 2015), pp. 1–15.
14. Baum, M., Dibbelt, J., Hübschle-Schneider, L., Pajor, T., & Wagner, D. (2014). *Speed-consumption tradeoff for electric vehicle route planning*. Proceedings of the 14th workshop on algorithmic approaches for transportation modeling, optimization, and systems (ATMOS 2014), pp. 138–151.
15. Baum, M., Dibbelt, J., Gemsa, A., Wagner, D., & Zündorf, T. (2015). *Shortest feasible paths with charging stops for battery electric vehicles*. Proceedings of the 23rd SIGSPATIAL international conference on advances in geographic information systems, pp. 44:1–44:10.
16. Cionini, A., D'Angelo, G., D'Emidio, M., Frigioni, D., Giannakopoulou, K., Paraskevopoulos, A., & Zaroliagis, C. (2017). Engineering graph-based models for dynamic timetable information systems. *Journal of Discrete Algorithms, 46*, 40–58.
17. Kontogiannis, S., Michalopoulos, G., Papastavrou, G., Paraskevopoulos, A., Wagner, D., & Zaroliagis, C. (2015). *Analysis and experimental evaluation of time-dependent distance oracles*. Proceedings of the 17th workshop on algorithm engineering and experiments (ALENEX 2015), pp. 147–158.
18. Kontogiannis, S., Michalopoulos, G., Papastavrou, G., Paraskevopoulos, A., Wagner, D., & Zaroliagis, C. (2016). *Engineering oracles for time-dependent road networks*. Proceedings of the 18th Workshop on Algorithm Engineering and Experiments (ALENEX 2016), pp. 1–14.
19. Gavalas, D., Konstantopoulos, C., & Pantziou, G. (2016). Design and management of vehicle-sharing systems: A survey of algorithmic approaches. In *Smart cities and homes* (pp. 261–289). Morgan Kaufmann, US.

20. Kypriadis, D., Pantziou, G., Konstantopoulos, C., & Gavalas, D. (2018). Minimum walking static repositioning in free-floating electric car-sharing systems. In *2018 21st international conference on intelligent transportation systems (ITSC)* (pp. 1540–1545). IEEE.
21. Gavalas, D., Giannakopoulou, K., Kasapakis, V., Kehagias, D., Konstantopoulos, C., Kontogiannis, S., Kypriadis, D., Pantziou, G., Paraskevopoulos, A. & Zaroliagis, C. (2018). *Renewable mobility in smart cities: Cloud-based services*. Proceedings of the 2018 IEEE symposium on computers and communications (ISCC 2018), pp. 1280–1285.
22. Gavalas, D., Kasapakis, V., & Dzardanova, E. (2019). Mobile persuasive applications. In *Encyclopedia of computer graphics and games*. Springer Science, in press.

Intelligent System for Predicting Motorcycle Accident by Reaching into a Smart City Using a Kriging Model to Achieve Its Prevention and the Reduction of Deaths in the Medium Term

Alberto Ochoa-Zezzatti, Brian Urrea, José Mejía, and Liliana Avelar

1 Introduction

In Juarez City, the traffic accidents are very common, like in any big city worldwide. Even though traffic accidents are most of the time a mixture of many kinds of motor vehicles, motorcycle accidents use to be more disastrous. Accidents in the road that involve motorcycles in them are around 1–3% of the total of the accidents; however, the perception of drivers of other type of vehicles toward motorcycle drivers, the nature and the control of the drivers over their vehicles, and the environmental conditions make the motorcycle driver especially vulnerable to accidents. These conditions make the motorcycle drivers twice likely to be seriously injured and 16 times more likely than four-wheeled vehicle drivers and passengers. Also, they represent 14% of the total losses from traffic accidents. Many reasons are involved behind this, for example, the problems that drivers have when they are turning or bending [1]. Also, more experienced car drivers are not used to spot motorcycles in their way, as well that motorcycle drivers are very likely to make risky moves in the road, not taking into account this detail about other vehicle drivers [2]. Fig. 1 shows the kind of data to be analyzed.

2 Related Research

In [3] it is shown that efforts have been made in Singapore to support the rearrangement and distribution of bicycles. They argue that for this type of system

A. Ochoa-Zezzatti (✉) · B. Urrea · J. Mejía · L. Avelar
Juarez City University, Ciudad Juarez, Mexico
e-mail: alberto.ochoa@uacj.mx

M. M. Banat, S. Paiva (eds.), *Smart Technologies for Smart Cities*, EAI/Springer Innovations in Communication and Computing, https://doi.org/10.1007/978-3-030-39986-3_8

DIASEMANA	URBANA	SUBURBANA	TIPACCID	AUTOMOVIL	CAMPASAJ	MICROBUS	PASCAMION	OMNIBUS	TRANVIA	CAMIONETA	CAMION	TRACTOR	FERROCARRI	MOTOCICLET
Domingo	Accidente en	Sin accidente	Colisión con	1	0	0	0	0	0	0	0	0	0	1
Viernes	Accidente en	Sin accidente	Colisión con	1	0	0	0	0	0	0	0	0	0	1
Martes	Accidente en	Sin accidente	Colisión con	1	0	0	0	0	0	1	0	0	0	1
Martes	Accidente en	Sin accidente	Colisión con	1	0	0	0	0	0	0	0	0	0	1
Sábado	Accidente en	Sin accidente	Colisión con	1	0	0	0	0	0	0	0	0	0	1
Sábado	Accidente en	Sin accidente	Colisión con	1	0	0	0	0	0	0	0	0	0	1
Sábado	Accidente en	Sin accidente	Colisión con	1	0	0	0	0	0	0	0	0	0	1
Miércoles	Accidente en	Sin accidente	Colisión con	1	0	0	0	0	0	0	0	0	0	1
Miércoles	Accidente en	Sin accidente	Colisión con	1	0	0	0	0	0	0	0	0	0	1
Jueves	Accidente en	Sin accidente	Colisión con	1	0	0	0	0	0	0	0	0	0	1
Lunes	Accidente en	Sin accidente	Colisión con	1	0	0	0	0	0	0	0	0	0	1
Lunes	Accidente en	Sin accidente	Colisión con	0	1	0	0	0	0	0	0	0	0	1
Lunes	Accidente en	Sin accidente	Colisión con	1	0	0	0	0	0	0	0	0	0	1
Viernes	Accidente en	Sin accidente	Colisión con	0	0	0	0	0	0	1	0	0	0	1
Viernes	Accidente en	Sin accidente	Colisión con	1	0	0	0	0	0	0	0	0	0	1
Lunes	Accidente en	Sin accidente	Colisión con	1	0	0	0	0	0	0	0	0	0	1
Jueves	Accidente en	Sin accidente	Colisión con	0	0	0	0	0	0	1	0	0	0	1
Miércoles	Accidente en	Sin accidente	Colisión con	0	0	0	0	0	0	1	0	0	0	1
Miércoles	Accidente en	Sin accidente	Colisión con	0	0	0	0	0	0	1	0	0	0	1
Domingo	Accidente en	Sin accidente	Colisión con	0	0	0	0	1	0	0	0	0	0	1
Viernes	Accidente en	Sin accidente	Colisión con	1	0	0	0	0	0	0	0	0	0	1
Martes	Accidente en	Sin accidente	Colisión con	1	0	0	0	0	0	0	0	0	0	1
Martes	Accidente en	Sin accidente	Colisión con	1	0	0	0	0	0	0	0	0	0	1
Martes	Accidente en	Sin accidente	Colisión con	1	0	0	0	0	0	0	0	0	0	1

Fig. 1 Data associated with traffic accidents in Juarez City

(SBC) to be useful, the correct number of bicycles should be deployed in the right places, as this affects the rate of use of bicycles and how bicycles circulate within the system. In [4], mechanisms based on multi-agent systems are proposed to identify the bicycle crossing points as well as better distribution of the same. A research was done by [5] where the problem of shared bicycles with ACO was discussed. In [6], open data applied in Washington DC to the system of shared bicycles with the purpose of locating the terminals according to the best satisfaction of the users.

3 Methodology Used

The kriging method can help solve problems using data from traffic history in the city and using all this data to find in which places it is more likely the accidents to happen. By this way, one can predict the conditions with which accident risk increases and how it can be resolved.

The kriging method is an algorithm that takes advantage from statistical data in maps. The method uses a mathematical formula that takes the data from points in a map and can predict the position of a new hypothetical point that is interpolated by the surrounding points, and in that way, it can give an accurate prediction of the said point, as is possibly seen in Eq. (1).

$$\hat{Z}\left(s_0\right) = \sum_{i=1}^{N} \lambda_i Z\left(s_i\right) \tag{1}$$

The above is the formula of kriging. The new point ($Z(s_0)$) is predicted by summing the surrounding N chosen points ($Z(s_i)$). The values measured are given an unknown weight (λ_i). This unknown weight is extracted from the distance of the point itself compared to the new point that is going to be calculated [7].

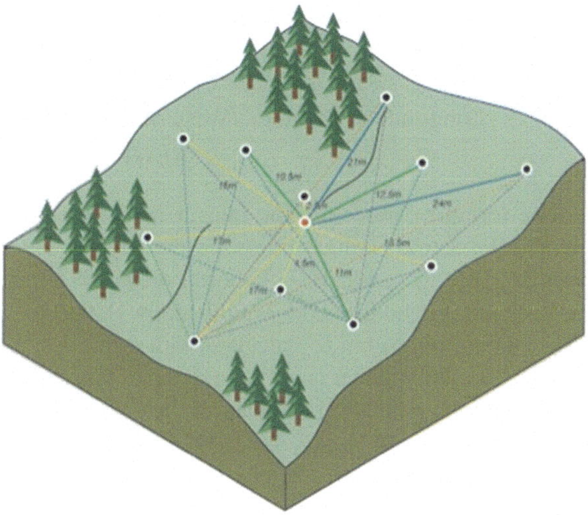

Fig. 2 A kriging model used to determine changes in an ecological model and its visual representation of a kriging model associated with the modeling of loss of forest in a landscape. (Source: http://desktop.arcgis.com/es/arcmap/10.3/tools/3d-analysttoolbox/how-kriging-works.htm)

We used a specialized software for our research, using QGIS (https://www.qgis.org/es/site/); the mapped points are used to calculate kriging and generate a special layer to show the results. This software is open source and comes very useful to create maps and get the calculations we need to get our results.

Mexico is the seventh-ranked country with the highest deaths caused by traffic accidents. There are around 2.5 million traffic accidents in the country. Ciudad Juarez is a city that has 1.32 million inhabitants who own in total 440,000 vehicles and that has between 4000 and 6000 accidents yearly, according to local statistic institutions. Juarez City is the fifth-largest city in Mexico and the third on deaths associated with traffic accidents, and for this reason, our research is based on both politics and advertising. The information is based on the number of deaths associated with traffic in the scope. Optimal for the users of said vehicles and as an affection in the same operation, shortening the negative effects of the falls after the accident evolving Italian scooters.

This data was provided to me to map them over the street map view of the city. We found that around 3000 total accidents in the city involve at least one motorcycle vehicle in them. Using a kriging model is possible to determine a correct tendency map in the future, as is shown in Fig. 2.

We fetched exclusively this data to map it in QGIS. In order to make all this information fit in the map in an ordered way, we took the spreadsheet file and randomly generated the latitude and longitude of the points of the map. All the other data fields were preserved as they came. Also, the spreadsheet software we used is the LibreOffice Calc, to generate the random positions needed to be limited in order for all the points to fit in the area of Juarez. Calc doesn't generate decimal point random numbers needed for the cartographic precision in QGIS. To solve this problem, we created an algorithm to create the positions. We used two different equations that represents each possible point in our model, as is possibly shown in Eqs. (2) and (3).

$$\frac{\text{Latitude} = \text{Random (Southernmost point, Northernmost point)}}{10,000} \tag{2}$$

$$\frac{\text{Longitude} = \text{Random (Westernmost point, Easternmost point)}}{10,000} \tag{3}$$

The formulas above are the integer random generator for the positions of the points. The actual cardinal limits for the city are in the order of the same integer number (31.5997 south, 31.7845 north, −106.3077 east, −106.5475 west). To solve the problem, we multiplied all numbers by 10,000 to open the range and to let the Calc random algorithm to calculate with a more flexible grid the position of the points. After the number is calculated, the number is divided by 10,000 again.

Once all the points were calculated, the spreadsheet file was parsed as a comma-separated values (CSV) file, so the QGIS software can read the data of the records and print the layer of the locations. Once this was done, kriging method was applied. QGIS has a method to perform this by itself. The kriging method was applied to adjust the grid and the area size and remove the OSM beneath, so the new calculated layer would not be too big. As they are being still too small, indicated by the absolute positions of the points, the new kriging calculated layer was difficult to adjust. We created a new project to add the new point layer. The points were static when imported from the CSV file, so a new layer, derived from this one, was created in order to freely move the points. QGIS has a plug-in that allows to load an Open Streets Map (OPM) view, so this one was loaded beneath the point layer. Then the points were manually positioned to the closest road one by one. Also, some of the points were in zones that compared to OPM were outside the area of Juarez. These were allocated inside the desired area and sometimes deleted, as is shown in Fig. 3.

Once this was done, kriging method was applied. QGIS has a method to perform this by itself. The kriging method was applied adjusting the grid, the area size, removing the OSM beneath, so the new calculated layer would not be too big. As they are being still too small, indicated by the absolute positions of the points. The new kriging calculated layer was difficult to adjust, as is possibly seen in Fig. 4.

If we compare this map above with the OSM, we can see the areas where the motorcycle accidents are more frequent. For instance, the area near Pronaf in the north is one of them. Knowing the city one can agree since that specific area has a lot of traffic during weekdays and also is the area where many universities are so many students ride motorcycles due to economic reasons. Also places near the airport (Abraham Gonzalez International) are very prone to motorcycle crashes. There must be a lot of traffic due to people arriving and requesting transportation, like buses, taxis, and Uber cars. To the south of this place there are neighborhoods and people go there constantly crowding the streets and the main avenue (Tecnologico Ave.).

In addition, we are proposing a tool for decision-making using ubiquitous computing to help parents in locating children during the tour in case they had forgotten some object that is required for the class, as you can see in Fig. 5.

Fig. 3 Results of each point in a possible traffic accident in Juarez City

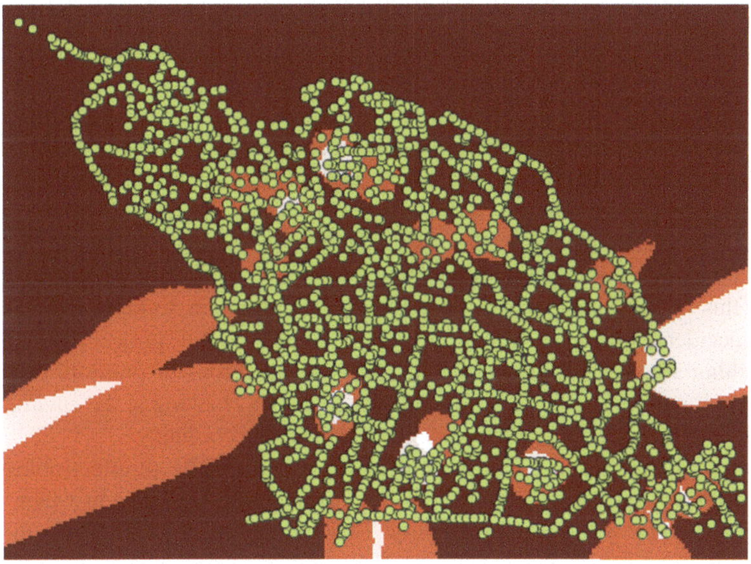

Fig. 4 Final analysis of each point where traffic accident evolving deaths associated with an Italian scoter will occur. The layer beneath the yellow dots (reddish colors) is the kriging-generated layer that shows the most probable places in lighter colors. Some places outside the city were shaded in light; these could be errors in the algorithm

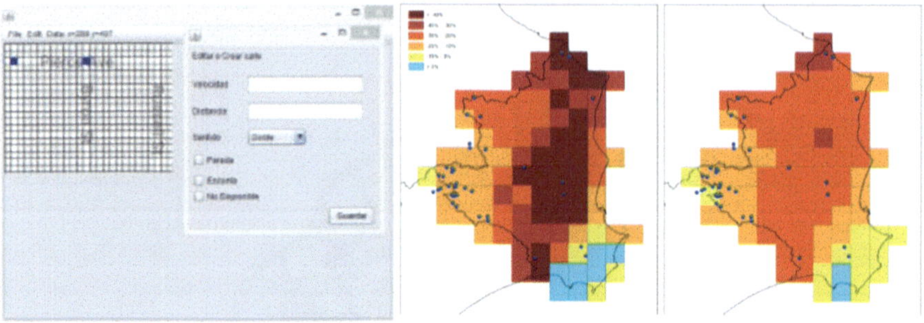

Fig. 5 Representation of each node and their visualization in the streets related with our kriging model

In the Google Maps city screenshot of traffic in Friday in the afternoon, one can see that the places in dark red, red, and orange are the places where traffic is worse than the areas in green, which we can say are more fluid and empty [8]. If one compares the areas in reddish with the kriging prediction in the Fig. 4, for example, the already mentioned Pronaf section and also the crossing bridge to neighboring city, El Paso, which very crowded due to security protocols in the United States. In Fig. 6, we showed all avenues where the majority of traffic accidents related to Italian scooters occur.

Traffic is not directly predicted by the kriging applied to the map, but the thing with traffic is that it is directly related to accidents. Many more vehicles in the streets lead to a higher chance to be hit by a car. Anxiety in drivers as a consequence of the jams will cause some of them to take risky decisions like passing a red light, overtaking the cars in front, and changing lanes, which might lead to traffic accidents and therefore worse traffic. Also, as mentioned before, motorcycle drivers are in more danger, because in jams, they use to go in between lanes, which makes them vulnerable to other vehicles shifting lanes.

All this information can provide very useful information, to know the places in which accidents may happen more frequently. Thanks to this, projects can arise on researching what causes this places to be more dangerous or prone to accidents or research on security helmets for motorcycle drivers, looking at the environment surrounding those places and building the equipment accordingly.

Equipment like helmets can prevent fatality to up to 28% of effectiveness [9]. But the effectiveness can be improved by analyzing the scenarios in which these accidents happen with more probability. This kriging applied to maps is the first step to this direction, because it closes down the investigation and study to smaller, denser areas in which the team can work on looking at possible reasons why this happens more often.

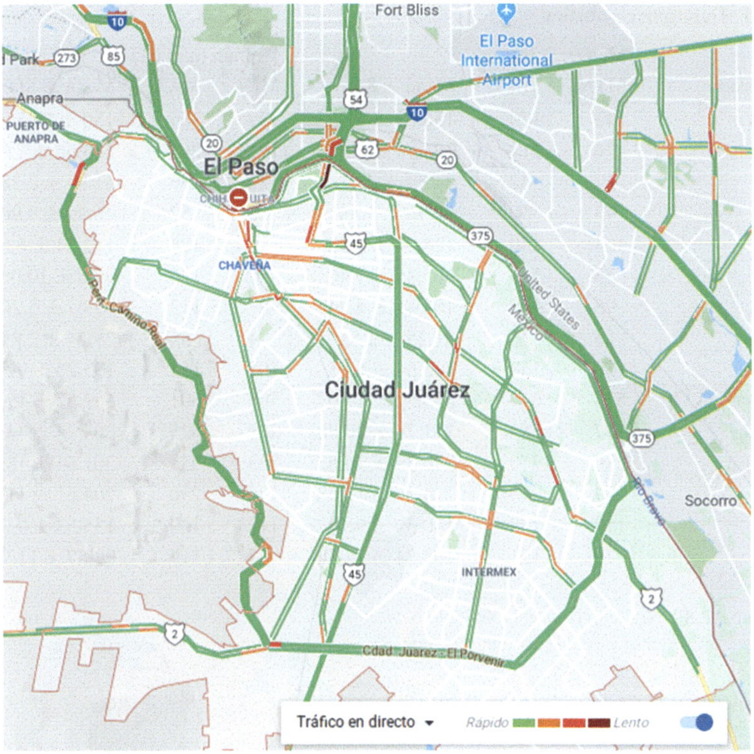

Fig. 6 Google Maps traffic map over Juarez area [8]. The green colors are fluid, and orange, red, and dark red are more jammed sites

4 Some Common Mistakes

As part of the strategy of the present research to address the above problem, a port is taken as a point of reference (52) of 86. The geographical location of this point is Avenida 16 de Septiembre (Alcalde) and Avenida Juárez (Avenida Vallarta), which specify the point where the depot is shown. This port was chosen for being the most demanded of all the SBC according to their number of trips. According to [10], the total number of bicycles for this point is 16, while for other ports the total is between 8 and 10, according to their point of location and demand. Table 1 shows the mean, minimum, maximum, and standard deviation of the collected data for men and women.

Table 1 Descriptive statistics

	Men ($n = 69$)			Women ($n = 61$)		
Dimension	MTN	MAX	Mean ± SD	MIN	MAX	Mean ± SD
Eu-Eu	142	170	156.62 ± 6.13	139	166	150 ± 5.65
G-Op	174	219	194.70 ± 9.05	171	201	185.66 ± 7.05
Al-Al	27	47	34.59 ± 3.02	26	36	31.97 ± 2.28
N-Ns	43	59	51.96 ± 3.34	38	57	46.92 ± 3.90
N-Gn	103	140	124 04 ± 6.93	99	128	112 54 ± 6.68
Ft-Ft	75	125	105.35 ± 9.46	77	126	99.2 ± 10.08
Ex-Ex	95	115	104 41 ± 4.34	87	112	99.67 ± 5.20
En-En	26	36	31.25 ± 2.55	24	35	29.34 ± 2.85
Zy-Zy	130	158	142.54 ± 6.85	105	154	133.26 ± 9.57
Perimeter G-Op	534	614	573.52 ± 17.07	508	601	552.57 ± 17.21
G-V-Op	277	383	321.78 ± 20.86	261	525	303.74 ± 34.70
Eu-V-Eu	286	370	320.29 ± 16.53	263	338	301.64 ± 18.01
V-Gn	197	249	224.39 ± 11.68	185	230	204.61 ± 11.08
V-Po	119	158	140.39 ± 9.14	118	160	135.53 ± 9.38
V-N	79	151	103.54 ± 14.30	67	119	92.57 ± 11.95

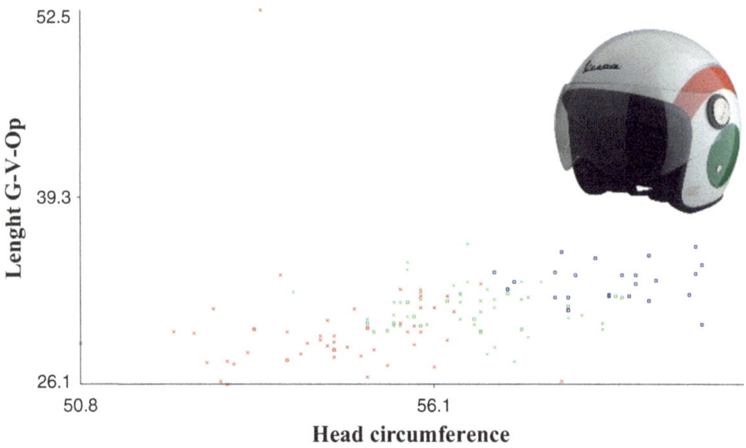

Fig. 7 Data mining analysis of our sample population to propose the ideal and optimal helmet to a shared bike system

5 Data Mining and Clustering Analysis

A clustering was performed on the complete database to find distinctions between people. The algorithm, on a large scale, detected three groups. When doing the comparison with the class gender, there is an error of 35.38% in the classification of the instances; however, the clusters reveal different characteristics of people (see Fig. 7).

6 Algorithm Classification

The algorithms tested do not differ a change of significance of 5%. As a result, it is not possible to indicate which algorithm performs better. However, with the results provided, it is observed that the attributes are informative. This after observing that they are better than applying the baseline classifier Zero R, which, part of trivial assumptions.

Identify data set. At this point, the information analysis about the location of the busiest points and limits that cover the route as well as bicycles per point is performed. Count trips per users. Once the set of data to be analyzed has been identified, a count is made on the users who go to what point to determine the most requested and at what time. Create instance. When the instance is created, it is necessary to give it a format as shown in Table 2. Prior to the execution of the algorithm, it is necessary to have an Id (node/ant), X (position x in the plane), Y (position y in the plane), and D (point-to-point loading). Tracing the route is possible to identify a complete network in which the routes made by the units are long. As part of the strategy, the points that make up the route are identified, which are shown in Fig. 8.

Table 2 Algorithm classification

Percent correct	Kappa statistic	Mean absolute error	RMS error	Relative absolute error	Root relative squared error	F measure	MCC	Area, under ROC
0.83	0.66	0.17	0.38	0.34	0.76	0.84	0.67	0.92
0.06	0.13	0.06	0.08	0.12	0.02	0.05	0.13	0.05
J48								
0.76	0.51	0.25	0.47	0.50	0.95	0.75	0.52	0.78
0.07	0.14	0.07	0.08	0.14	0.15	0.08	0.14	0.09
Logistic								
0.81	0.63	0.21	0.40	0.42	0.79	0.83	0.63	0.87
0.09	0.17	0.06	0.09	0.13	0.17	0.08	0.17	0.07
OneR								
0.77	0.54	0.23	0.47	0.45	0.95	0.79	0.55	0.77
0.06	0.12	0.06	0.06	0.12	0.12	0.05	0.12	0.06
ZeroR								
0.53	0.00	0.50	0.50	1.00	1.00	0.69	NaN	0.50
0.01	0.00	0.00	0.00	0.00	0.00	0.01	NaN	0.00
IBk								
0.80	0.58	0.22	0.39	0.44	0.78	0.83	0.60	0.85
0.05	0.10	0.04	3.05	0.08	0.10	0.04	0.09	0.05
PART								
0.74	0.48	0.26	0.49	0.53	0.99	0.74	0.49	0.76
0.80	0.15	0.07	0.08	0.15	0.17	0.08	0.15	0.08

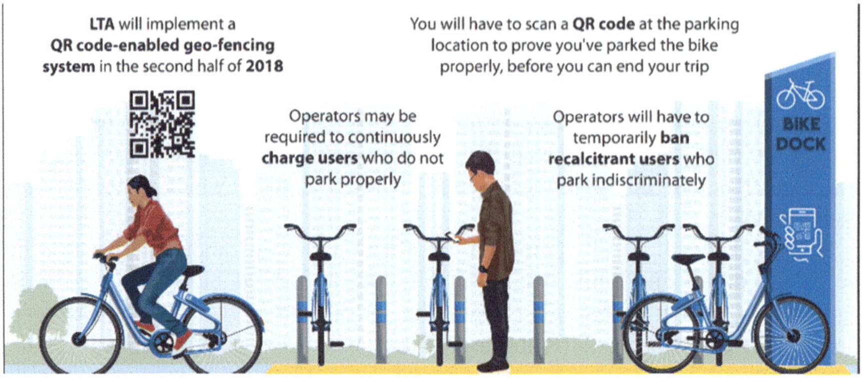

Fig. 8 Conceptual model of the correct use of bikes in a shared system

7 Results

When the data-mining tools are implemented in high-performance parallel processing systems, one can analyze massive data bases in just few minutes. Faster processing means that the users can automatically experiment with more models to understand complex data [3]. High speed is practical for the user and makes possible to analyze immense amounts of data. The great data bases, as well, can produce better predictions. The data bases can be huge on depth as well as on width. More columns. So, many times, analysts must limit the number of variables to examine when manual analyses are done due to limitations on time. However, variables are suppressed because they seem to be without importance and can provide information about unknown models. A high-performance data mining allows users to explore the whole data base, without a set of variables preselection. More rows determine most ideal use of helmet in Italian scooters. Bigger samples produce less estimation errors and deflections and allow users to make inferences about small but important population segments. To improve our proposal, we decided to specify the spaces assigned to the shared bicycle system to be able to assure the prompt mobility of each issue associated with a bicycle and its user, as can be seen in Fig. 9.

8 Conclusions

If we compare this map above with the OSM, we can see the areas where the motorcycle accidents are more frequent. For instance, the area near Pronaf in the north is one of them. Knowing the city, one can agree, since that specific area has a lot of traffic during weekdays and also is the area where many universities are, so many students ride motorcycles due to economic reasons.

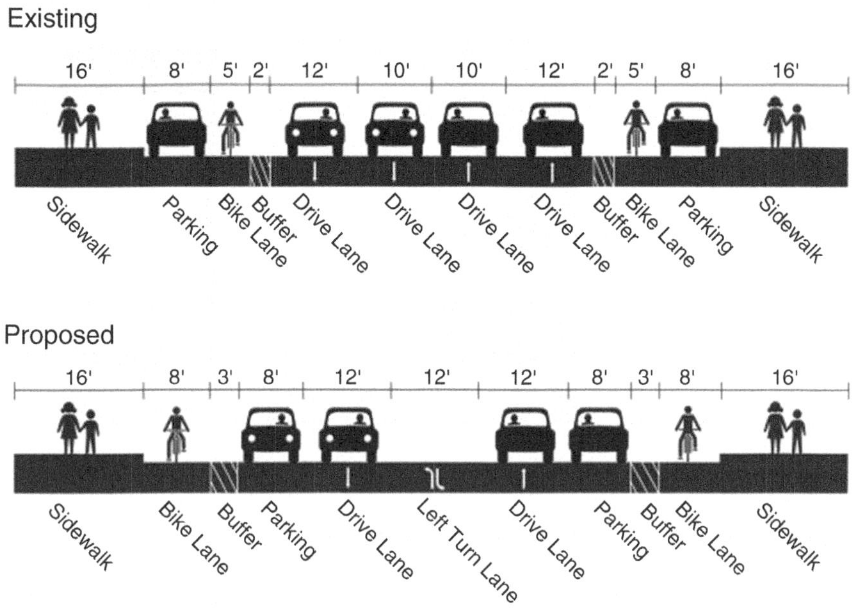

Fig. 9 Distribution of an adequate model to separate spaces associated with a shared bicycle model

A relevant aspect of the present investigation is to adapt the public policies associated with deaths due to traffic accidents and to determine the effect that can be imposed by driving restrictions in conditions of abuse of alcohol or other illicit substances.

Of the results obtained in the present research, the conjugation of the matrices and their symbolic capital is achieved in order to be able to apply the data in an algorithm to find the best routes. It is possible to conclude that the use of the intelligent algorithm and the kriging model are the shortest route with the most significant value in the whole amusement park or a Smart City. The main results found, it is possible to conclude that the use of the algorithm the shortest route with the most significant value in the whole amusement park, finding that if you visit 11 of the main attractions you get to have a unique experience with the maximum heat, in this case of 88.7 and the most important attractions from the arrival.

With these results it is possible to solve the problem raised in relation to the limited times of users and visitors in journeys and scheduled times, if the entire Smart City is not known if it is possible to know the most significant and symbolic. With the use of this innovative application, using the Dijkstra algorithm associated with a kriging model can be applied on a mobile device to determine the most important places where it is possible to improve a trip as is analyzed in [3, 8–10]. The route of the main attractions is proposed in a smart 3D screen used in social networks, with real-time information in a mobile device, as is proposed in Fig. 7,

Fig. 10 We determine the use of an intelligent application to determine the relevance and ranking of adequate selection of a travel in bicycle in a Smart City

and the establishment of a trip, including saving money and time and enjoying with the rest of the community. The most important contribution of the present research is the use of the algorithm in an amusement park where it includes the real-time and the symbolic capacity in a route and that allows to avoid the expense of time and money in this type of trips in a Smart City.

After the experimentation of the present research, it is observed that, through the application of computational techniques, it is possible to improve a SBC in terms of distance and time and derived from it is possible to keep the units in good condition. Using an intelligent application, we can determine through Internet of Things (IoT) the requirements of each user and specify the actions to be carried out in order to use a bicycle or electric bicycle, and if the distance is greater or must be transshipped, it also adequately determines the distance between transshipments as is analyzed in [11], and the way to obtain shorter waiting times is efficient, as can be seen in Fig. 10.

9 Future Research

Is recommended for the application in a mobile device with current photographs and the main recommendations in the time available for the trip. The intelligent tool will be used by different types of people who need to travel together with different situations related to displacement. In addition, this application will be used as a recommendation system when traveling to other cities or places in different societies and will explain different scenarios based on time, limited resources, and location. Another field topic that will benefit from a more appropriate organization is logistics of the product or service as it describes the use of cultural algorithms to improve a logistics network. We propose the use of a mobile computing device that can

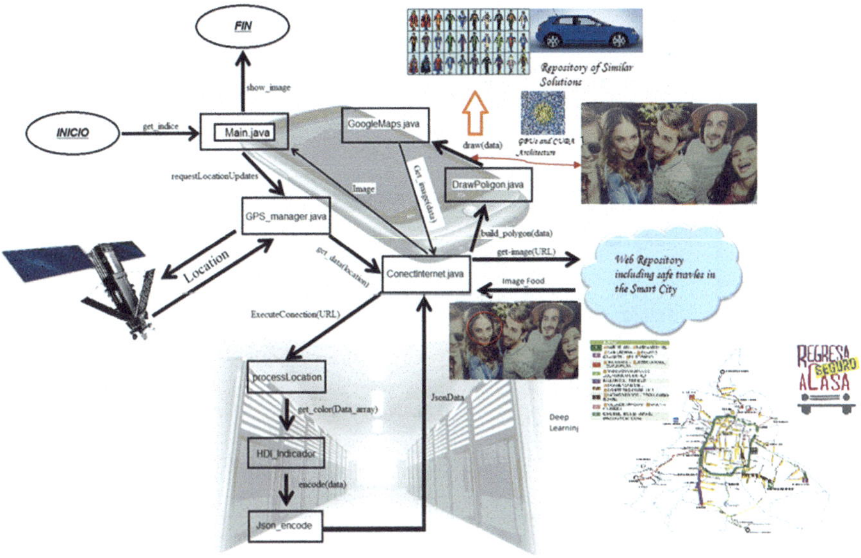

Fig. 11 Representation of the model associated with the reduction of deaths in nighttime collective trips associated with alcohol in young people in a Smart City

indicate during the tour fun options based on our own tastes and priorities, according to the type of activity you want to perform.

We intend to perform more tests on other equipment with different characteristics, either hardware with a graphics card with higher processing cores and a webcam with better resolution or other software versions in order to obtain better results in the quality of the processed video using diverse applied methodology used by kriging as in [4–6, 11, 12]. This platform is also intended to be multiplatform for functions not only in Linux but also in other operating systems, whether it is the Windows or Mac environment. One of the main problems in a Smart City are accidents with more than one death associated with reckless driving due to alcohol, we propose that using a mobile device and ubiquitous computing techniques determine through photos of the online status of a group of friends that they will travel together, in order to determine who through the recognition of the iris could be the designated driver, something crucial for the lives of the group, as can be seen in Fig. 11.

One of the big challenges in a Smart City will be that the young population is becoming groutier, and therefore, their outings are nocturnal and last longer, and this means more alcoholic time and therefore with a higher incidence in deaths due to vehicular accidents in a Smart City.

As a future work, it is proposed to submit the SBC instance to solve the Capacitated Vehicle Routing Problem (CVRP) in a parallel environment in order to observe the behavior of both schemes, that is, sequential and parallel. In the near future, making trips of greater distance involves several aspects to be solved, among

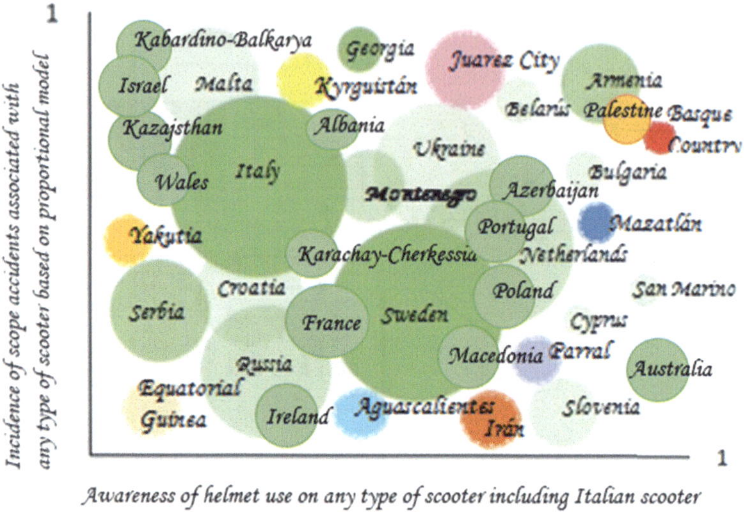

Fig. 12 Comparison of transit accidents, in 39 different societies, relates with any kind of vehicle or motorbike including Italian scooter: green, European and Australian societies; yellow and orange, Asian societies; brown, African society; blue and purple, Mexican societies; and Pink, our border society of this study

them mainly the safety of the users through optimal and ideal helmets, a smart grid to help with the energy that the electric bicycles require, a distribution model of space for the Smart City, and finally an interconnected system that manages to reduce the impacts by scope among the various vehicles, and that is why it is proposed to determine how to improve this group of actions in a Smart City in the long term. In addition to our future research, it will be specified in a serious game that allows the use of avatars as it should be correctly used helmets proposed in a Smart City and the future impact in order to reduce by 2027 the high incidence of deaths by reach in this border society [13]. The excuse in young people do not use halmet is that in an Italian scooter they feel like the wind and have not wings but dying to fly to realize all your dreams, they will not matter if you will never try to begin and have many reasons because we are young to promote the use of this kind of helmet [14]. We are young to not use a specific helmet as in Wales and Kazakhstan, as is shown in Fig. 12.

References

1. Department of Transport. (2004). *In-depth study of motorcycle accidents*. Londres: University of Nottingham.
2. Ouellet, J. (1981). *Environmental hazards in motorcycle accidents*, ResearchGate (no. 1, pp. 117–129).

3. Ma, F., Han, L., Zhou, Y., Chen, S., & Pu, Y. (2018). Multi-island genetic algorithm and kriging model-based design of vehicle product comprising multi-material. *IEEE Access, 6*, 53397–53408.
4. Crosby, H., Damoulas, T., Caton, A., Davis, P., Porto de Albuquerque, J., & Jarvis, S. A. (2018). Road distance and travel time for an improved house price kriging predictor. *Geo-Spatial Information Science, 21*(3), 185–194.
5. Zhang, Y., Zheng, X., Wang, Z., Ai, G., & Huang, Q. (2018). Implementation of a parallel GPU-based space-time kriging framework. *ISPRS International Journal of Geo-Information, 7*(5), 193.
6. Zhang, X., Zuo, W., Zhao, S., Jiang, L., Chen, L., & Zhu, Y. (2018). Uncertainty in upscaling in situ soil moisture observations to multiscale pixel estimations with kriging at the field level. *ISPRS International Journal of Geo-Information, 7*(1), 33.
7. Cómo funciona Kriging—Ayuda | ArcGIS Desktop, Pro.arcgis.com. (2019). [Online]. Available: https://pro.arcgis.com/es/pro-app/tool-reference/3d-analyst/how-kriging-works.htm. Accessed 7 Mar 2019.
8. Liu, D., Wu, J., Lin, K., & Wu, M. (2019). Planning of multi energy-type micro energy grid based on improved kriging model. *IEEE Access, 7*, 14569–14580.
9. Cortesi, A. F., Jannoun, G., & Congedo, P. M. (2019). Kriging-sparse polynomial dimensional decomposition surrogate model with adaptive refinement. *Journal of Computational Physics, 380*, 212–242.
10. Mazumdar, A., Chugh, T., Miettinen, K., & López-Ibáñez, M. (2019). On dealing with uncertainties from kriging models in offline data-driven evolutionary multiobjective optimization. In *International Conference on Evolutionary Multi-Criterion Optimization* (pp. 463–474). Springer, Cham.
11. Wang, B., & Hu, J. (2018). Some monotonicity results for stochastic kriging metamodels in sequential settings. *INFORMS Journal on Computing, 30*(2), 278–294.
12. Bostan, P. A., Heuvelink, G. B., & Akyurek, S. Z. (2012). Comparison of regression and kriging techniques for mapping the average annual precipitation of Turkey. *International Journal of Applied Earth Observation and Geoinformation, 19*, 115–126.
13. Bortner, C. W., et al. (2015). *Ant colony optimization applied to the bike sharing problem*. Ph.D. Thesis, Carnegie Mellon University.
14. Chen, L., et al. (2015). Bike sharing station placement leveraging heterogeneous urban open data. In *Proceedings of the 2015 ACM International Joint Conference on Pervasive and Ubiquitous Computing* (pp. 571–575).

Development of a Java Library to Solve the School Bus Routing Problem

Alberto Ochoa-Zezzatti, Ulises Carbajal, Oscar Castillo, José Mejía, Gilberto Rivera, and Saúl Gonzalez

1 Introduction

Each day, buses make two trips. In the morning, students are picked up at their homes, while in the afternoon, buses transport students from the school to their homes [3]. However, bus stops are not generated by area, where the number of students is concentrated, and thus, reduce the number of stops that would be made by improving the transfer time. There is an additional generalization to the problem, which is to allow transfers, that is, students who attend different schools can share a single bus and change it during their route to school [5]. However, the waiting time is linked to the number of transfers, the more transfers as the time increases. The authors of the study chose to solve the SBRP by means of an algorithm of optimization of colony of ants, because they were developed to find good solutions, in a time of reasonable calculation for routing problems [3], which was used to solve the routing phase, in order to find the best route. When analyzing the results, it was observed that some parameters have greater impact than others and that the distance of the routes can be reduced, but this would increase the time; the problem was solved by a two-phase scheme where the place where they would be assigned was assigned. Students were collected, and the second phase consisted of the optimization of the routes, through the algorithm of colony optimization of ants, and obtained an improvement of 15.2% in the cost and in the time of travel as well

A. Ochoa-Zezzatti (✉) · U. Carbajal · J. Mejía · G. Rivera · S. Gonzalez
Universidad Autónoma de Ciudad Juárez, Ciudad Juárez, Mexico
e-mail: alberto.ochoa@uacj.mx

O. Castillo
Instituto Tecnologico de Tijuana, Tijuana, Mexico

M. M. Banat, S. Paiva (eds.), *Smart Technologies for Smart Cities*,
EAI/Springer Innovations in Communication and Computing,
https://doi.org/10.1007/978-3-030-39986-3_9

as to take them to class in the morning and back to their homes. An estimate of the distance traveled is made; for all buses, the requirements of the school are that students should not travel more than 25 kilometers [4].

1.1 Related Research

A combination of transport arrangements with different capacities is proposed [2]. Considering several means of transport at the same time (cars, vans, buses), it is observed that the routes are not generated, but the stops. The SBRP has a total of five subproblems, including data preparation, bus stop selection, bus route generation, school bell time adjustment, and route scheduling, which can be seen in Fig. 1.

These subproblems are closely related to each other and must be resolved in an integrated manner [5]. In each proposed study, you can find different restrictions, such as the distances between bus stops, the time of arrival to classes, and the capacity of the buses, applying the restrictions, and in a case where it was necessary to transport 519 students to a primary school, the number of buses from 26 to 18 was minimized, and the bus utilization average improved from 60.49% to 87.37% [1]. For this, an algorithm was developed that can solve part of the SBRP, which when applied can show the efficiency compared with manual methods and thus is able to reduce the distances traveled. A model is analyzed to optimize routes, by grouping people traveling with the same destination. Using the algorithm DBSCAN, this algorithm takes advantage of the density of the points to determine the groups; this indicates that a circle is defined around this, obtaining its neighbors. For the

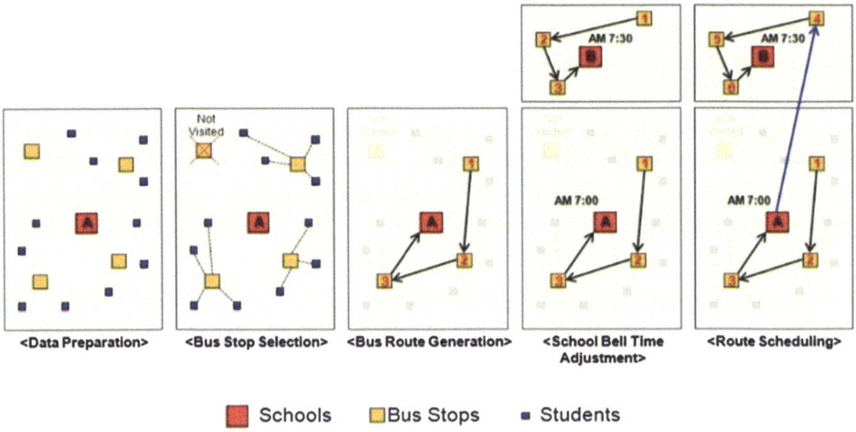

Fig. 1 Subproblems Involved school bus routing problem

existence of a cluster, it is necessary to have a minimum number of points within this cluster, that is, it is necessary to have a minimum number of neighbors of the point. Therefore, the algorithm iterates through the data set, verifies dense regions, and associates the neighboring points with the same clusters, taking into account the condition of the minimum number of points per group [2]. The algorithm considers the density of the points; sometimes the minimum number of clusters is not counted, so these points are taken as noise. Tests were conducted with a population of 437 people, where the DBSCAN algorithm generated 33 clusters for a total of 394 people representing 90.2% of the total; the rest of the population was taken as noise; this is due to the low density in the area which they are from. The results obtained are satisfactory for the authors, since they can verify that the time of transfer of the people can be reduced, and, thus, reduce the traffic of the city, to reduce the number of vehicles on the roads. Several studies have been analyzed with respect to the School Bus Routing Problem, where each of them has been implemented in different environments, with different restrictions, such as distance and time, or having as objective only the creation of bus stops. Not all show the improvements and optimizations that could be obtained by applying the algorithm, and some of them are not developed as a program, and a few are not easy to implement in other environments, since they are designed for a specific environment. It is proposed to generate the routes in an optimized way, and only buses will be used, since it is focused on a school district. It can be observed that similar works have already been implemented satisfactorily; with this basis, it is considered possible to implement a similar tool for Juarez City.

1.2 Problem Statement

The SBRP has been studied constantly since its first publication by the authors Rita M. Newton and Warren H. Thomas in 1969. Since most of the studies have been carried out thanks to the appearance of the day-to-day problems with their own restrictions, we do not have an overview of the SBRP [13]. The SBRP can be solved in five steps as already mentioned, which are data preparation, selection of stops, route development, time adjustment, and route planning. Although these subproblems are not independent, but are highly interrelated, they are treated separately due to the complexity and size of the problem. Although the SBRP itself is a unique and independent problem, a single subproblem or a combination of its subproblems can be classified as a variant of existing optimization problems. By itself, the subproblem of generation of bus routes is very similar to the vehicle routing problem (VRP) since it seeks to generate efficient routes; it also has part of the traveling salesman problem (TSP) since in this they have to travel efficiently and through all cities; in SBRP, you have to go through all the bus stops.

2 Framework of Reference

A library programmed in Java was realized, with a basic scheme of a database that stores the coordinates of the school, the students, and the bus stops. The library was able to take the coordinates of the database, calculate the distance between stops, and find an optimal route between them. By making these processes work together, the library can be used to solve the SBRP regardless of the region in which the problem is located, that is, whether it is in the urban area or in a rural region. For the operation, you must have the data of the coordinates, as well as the school and those of the bus stops loaded in the database; these data are necessary for the proper functioning of the library. The library is designed to be used by calling two functions within it, getDatos and neighborC; with these functions, the information was extracted from the database, and the optimal route was calculated.

2.1 Smart Cities

It is a term that began to be used in the 1990s to refer to cities where, thanks to the integration of information and communication technologies (ICT), energy and transport have progressed [6]. They begin by modernizing their critical infrastructure to provide sustainable development, help improve the efficiency of available resources, and improve the quality of life of citizens [7]. Among the advantages presented by smart cities are improving the efficiency of public administrations, interconnecting the different systems that make up the cities, facilitating the management of a greater number of public services, reducing the traffic of the transportation systems, and improving the speed of response in emergencies [6]. Although there are multiple areas that benefit from the use of ICT within smart cities, this work is closely related to the area of public transport, given that Cd. Juarez currently does not have a technological implementation for the use of public transportation that serves citizens in their mobilization within it.

2.2 Graph Theory

The first step to solve the SBRP is the preparation of the data, which refers to having the roads, the geographical location of the school, and the students. As a sample of them, 90 points were taken to simulate the students. With the help of Google Maps, it was possible to create a map with the geographical location of the school and the students; also the coordinates of each point could be obtained. In Fig. 2, you can see the locations of the students in color and the school.

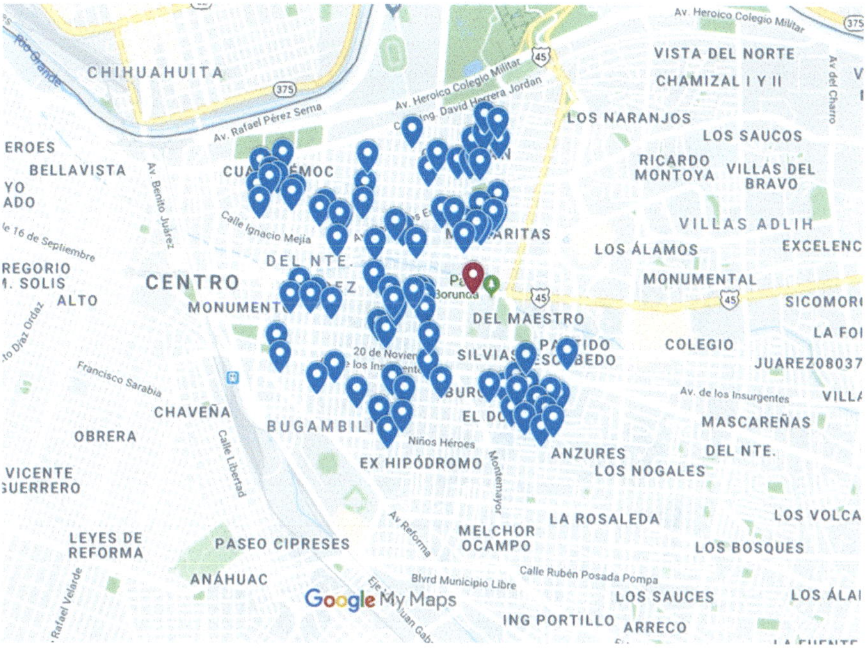

Fig. 2 Location of the points on the map

2.3 *Stop Selection*

By having the geographical location of the school and the students, we proceeded to select the bus stops that would be visited for the promotion and descent of the students. These do not have a specific distance between them; they were located in the points with student population, because it was proposed that students should not travel more than 500 meters of walking distance. In Fig. 3, the marks of the points are identified which were divided by color; the blue marks represent the location of the students, the green marks represent the locations of the bus stops, and the red mark is for the location of the high school.

Once the bus stops were created, groups were created, and these groups are made up of the bus stop and the students; to these groups, the students were assigned with the condition that they cannot cover a distance greater than 500 meters of walking distance. In the following Fig. 4, it is possible to see a part of the created groups.

Once the groups were created, the students were assigned to their corresponding stop; in this part, it was considered in addition to the distance that has to be traveled by the student and the radius of the zone that the group covers. For a student to be assigned to a group, he has to comply with the condition of not moving more than 500 meters to reach the stop; if the radius of two bus stops is at intersection and the student is between that intersection, it is assigned to the group you are closest to. In Fig. 10, the groups created, and the students assigned to them are shown (Fig. 5).

Fig. 3 Location of bus stops

Fig. 4 Creation of the groups to be analyzed

Fig. 5 Students assigned to groups

2.4 *Dijkstra's Algorithm*

This algorithm arises in order to find the shortest distance from a node of origin to any other specified node and was created in 1959 by Edsger Wybe Dijkstra. Each node is assigned a weight; this weight is a value that represents how short the distance of that node is at the beginning. This algorithm is considered to belong to the algorithm O(n^2) where n is the number of nodes belonging to the network, given that increasing the number of nodes usually increases the solution time [10]. It consists in broad strokes, having initialized the distances between nodes and grouping the nodes or vertices into three types, the nodes marked, adjacent, and distant. The initial node is a marked node, the adjacent nodes are nodes connected to a marked node, and the distant nodes are nodes that have not yet been taken into account. It must be taken into account that each adjacent node saves which node was accessed. The procedure is as follows: the initial node is a node marked as visited and with zero cost, the adjacent nodes are marked to the current node, and its cost is calculated from the current node; from this point, the algorithm will cycle until all the nodes are marked nodes; from the marked nodes, the lowest cost is taken and becomes the current node, and the nodes connected to the current node that were adjacent to another node become adjacent nodes of the current node if its cost from the current node is smaller than they already have [11].

2.5 *Technological Framework*

Since the data was stored, call the distance function, which receives the two lists of arrangements, and return a double-type matrix; in this function, the calculations of the distances between points were made, and the matrix that returns counts with the distances between said points. To calculate the distance between the points, two nested cycles and the following equation were used:

$$d = \sqrt{(X2 - X1)^2 + (Y2 - Y1)^2} \tag{1}$$

where the "*X*" is taken as the latitudes and the "*Y*" as the lengths; this is inside nested, to calculate the distance of a starting point toward all other stops; in this way, all the stops will be connected between them. When the distance is calculated with the same point, the diagonal of zero is generated, taking into account that the distance is the same in the inverse points, that is, the distance in the point (1,2) is the same distance in point (2.1), and it can be confirmed that the distance between the points is calculated correctly; in the following Table 1, the distances between points are shown.

2.5.1 vecinoC Function in Java

This function is where the calculation is performed to obtain the optimal route, and the function receives two double matrices which are the distance matrices; the reason why it receives twice the same matrix is due to the procedure that is performed to obtain the optimal route; when doing the search in the matrix, the values are being modified with 100.00. In this function, the near-neighbor algorithm is implemented, in which nested cycles are used; the first cycle is to have control of the initial point of each route and to look for different optimal paths from other points and to find a better route. The second and third cycle is where the search for the minimum value between each point of the matrix is made (Fig. 6).

Table 1 Distance matrix between points

1	0	1	2	3	4	5	6	7	8
0	0	0.004754	0.008854	0.016618	0.008528	0.013256	0.005438	0.009781	0.008884
1	0.004754	0	0.004237	0.015505	0.007914	0.01495	0.00897	0.013555	0.013046
2	0.008854	0.004237	0	0.014227	0.008267	0.016203	0.01211	0.016556	0.017278
3	0.016618	0.015505	0.014227	0	0.008102	0.009119	0.014198	0.015547	0.024567
4	0.008528	0.007914	0.008267	0.008102	0	0.007995	0.007189	0.010359	0.016796
5	0.013256	0.01495	0.016203	0.009119	0.007995	0	0.008442	0.00738	0.018732
6	0.005438	0.00897	0.01211	0.014198	0.007189	0.008442	0	0.004585	0.010616
7	0.009781	0.013555	0.016556	0.015547	0.010359	0.00738	0.004585	0	0.012128
8	0.008884	0.013046	0.017278	0.024567	0.016796	0.018732	0.010616	0.012128	0

Fig. 6 First cycles

```
for(int x=0; x<myList.length; x++)
{
   start=x;
   columna=x;
   arreglo[renglon][columna]=x;
    renglon++;
   ruta.clear();
   ruta.add(start);


   for(int i = 0; i <myList.length-1; i++)
       {

           for(int j = 0; j < myList.length; j++ )
           {

               if(min>myList[start][j])
               {
                       min=myList[start][j];
                       nodo=j;

               }

           }
```

Within the third cycle, there is a condition where the minimum value of the distance matrix was analyzed when finding it and the value of the node is taken to form the route. In the following image, the code of the cycles is shown. When leaving the third cycle, a sum was made with the distance between the nodes and another for where the matrix will be modified with the maximum value, to mark the points where the minimum value was found and discard them when analyzing the matrix again. The route was stored in an ArrayList; this is updated when leaving the cycle that modifies the matrix and when leaving the cycle for another condition in which the distances are compared, and the minor is conserved.

3 Development

In this section, the most outstanding aspects are presented during the development of the project; they are the diagram of the proposed solution, the methodology used, and the development process.

Table 2 Distance matrix with maximum diagonal

	0	1	2	3	4	5	6	7	8
0	100	0.004754	0.008854	0.016618	0.008528	0.013256	0.005438	0.009781	0.008884
1	0.004754	100	0.004237	0.015505	0.007914	0.01495	0.00897	0.013555	0.013046
2	0.008854	0.004237	100	0.014227	0.008267	0.016203	0.01211	0.016556	0.017278
3	0.016618	0.015505	0.014227	100	0.008102	0.009119	0.014198	0.015547	0.024567
4	0.008528	0.007914	0.008267	0.008102	100	0.007995	0.007189	0.010359	0.016796
5	0.013256	0.01495	0.016203	0.009119	0.007995	100	0.008442	0.00738	0.018732
6	0.005438	0.00897	0.01211	0.014198	0.007189	0.008442	100	0.004585	0.010616
7	0.009781	0.013555	0.016556	0.015547	0.010359	0.00738	0.004585	100	0.012128
8	0.008884	0.013046	0.017278	0.024567	0.016796	0.018732	0.010616	0.012128	100

3.1 Proposed Solution

3.1.1 Implementation of the Near-Neighbor Algorithm

As mentioned, this algorithm was implemented in the neighborC function; when this algorithm was performed in Java, the optimal route for the SBRP solution was obtained, and the implementation was carried out using the distance matrix looking for the minimum between the points. The algorithm is completely adaptable for the number of stops that are necessary; this is because the routes are stored in a matrix that takes the size of its lines and columns of the size of the ArrayList.

When calculating the distance of the equal points of the matrix, for example, m [0] [0] will be placed with a double greater than all the distances shown in Table 2; this modification was made in the distance function in order to discard these points as minimum since if they remain as zero they would be taken as small values. The following table shows how it was formed.

In this way, it is ensured that the minimum real values are taken, having the minimum of each point; two situations necessary for a good result are processed, and both situations happen when having the minimum value; one is when finding said value, and that value is substituted in the matrix by the greater value; this would mark the point as visited, and it would be discarded that that point be contemplated again; the next situation is to save the point of the visited matrix in an ArrayList which would be forming the optimal route.

3.2 Construction of the Prototype

The database has five tables, in which the information contained is necessary to resolve the SBRP. The tables are the following:

- Depósito_autobuses: This table has only two fields, since it is designed to have better control of the buses that count and its capacity; the fields are id_autobus;

this field is the field to identify buses; and the field ability is to store the amount of students who can ride the bus.

- Parada_autobus: This table is designed to store the location of the bus stops; this is the most important table in the design since from here the data is taken to perform distance calculations in the Java library: in the field id_stop, this field is to identify the bus stop; in the field id_autobus, the bus that would pass through that stop is associated; in the field students_paid, the number of studies that count by stop is stored; in the field lat_bs, this field saves the latitude of the bus stop; and in the field lon_bs, the length of the bus stop is stored.
- Detail_parado: This table is designed to have the relationship between student and stop, to have a better detail of it, and to have a control that the student corresponds to the bus stop; the field id_stop is the identifier of the stop table of bus, and just like the field id_stop, the field student_id is the identifier of the student table.
- School: This table has the basic data of the school, including the location; in the fields, that account is school_id, which is the one with which the school is identified, and the field name, latitude, and longitude coordinates are stored in the lat_esc and lon_esc.
- Student: In this table, the basic data of the students will be stored, such as name, paternal surname, maternal surname in their corresponding fields, and the coordinates for the location in the lat_est and lon_est fields (Fig. 7).

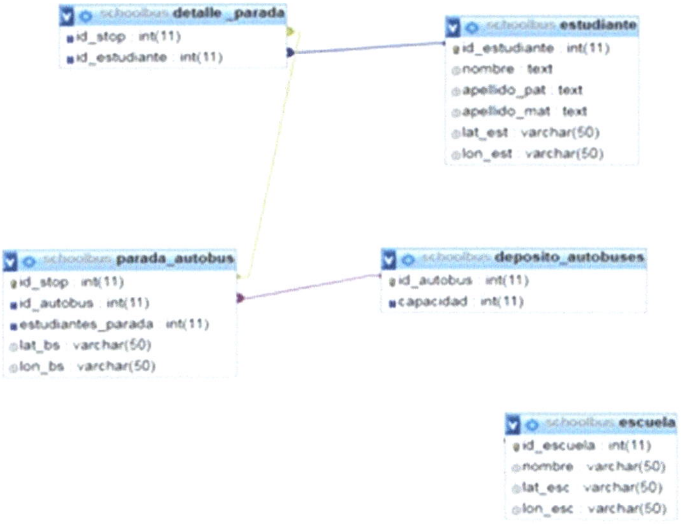

Fig. 7 Database relationship

```
public class conectar {
    private static Connection conn;
    private static final String driver = "com.mysql.jdbc.Driver"; //driver mysql-connector
    private static final String user="root";                      //Usuario de la bd
    private static final String password="";                      // psswd de la bd
    private static final String url="jdbc:mysql://localhost:3306/schoolbus";  //direccion de la bd
```

Fig. 8 Class connect

3.2.1 Data Connection and Management

For a better implementation of the code, a connection was made between the Java library and the database; with Java, this was possible by integrating a library called mysql-connector-java; in the project, version 5.1.46 was used. In this way, the connection can be handled as a class where the data of the database, such as the user, password, and location of the database, are passed as variables (Fig. 8).

Just as the class performs a function to call it from any other function within the program, taking advantage of that, it was sent to call from another class where the information obtained is processed, which are the coordinates of the bus stops. Being connected to the Java database, you can execute SQL query and extract the data from the stop_autobus table, since the information in the table could be processed in an array with the distances between the points.

3.3 Validation

A first route was obtained which starts at stop one, followed by two, four, six, seven, five, three, zero, and eight ending at the initial stop (1, 2, 4, 6, 7, 5, 3, 0, 8) which I throw a total distance of 7.9 kilometers. Each number on the route represents a bus stop, where zero was assigned to the school.

In Fig. 15, it shows how the route would be when following the route (Fig. 9).

4 Results

By adding the external cycle to the algorithm and performing the tests with the library, a new route was obtained for the graph, which would start at stop number four, following stops six, seven, five, three, two, one, zero, and eight, ending with the initial stop (4, 6, 7, 5, 3, 2, 1, 0, 8); the total distance of this route is 7.7 kilometers. It is possible to have a better route, since, when starting from a different point, the minimum distances to another bus stop change, or they are not discarded because it is a stop already visited. In Fig. 10, the route is shown following the route that was obtained.

Fig. 9 First route obtained

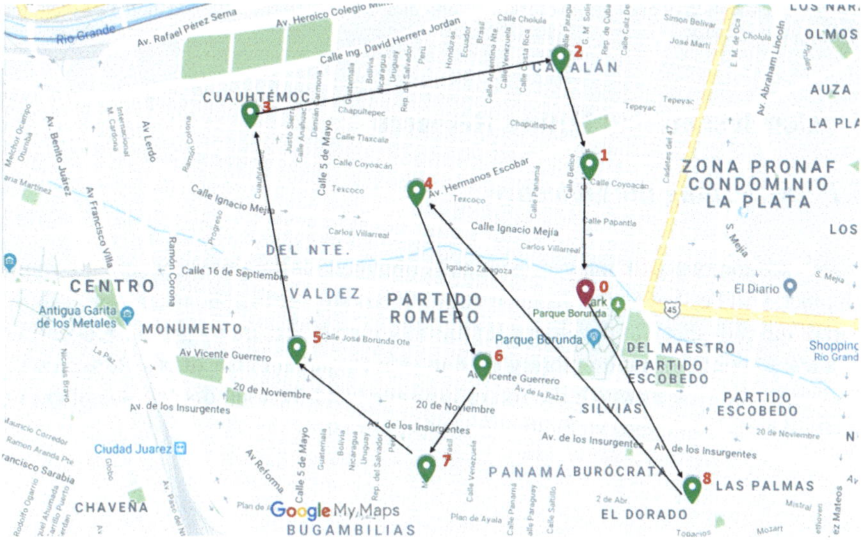

Fig. 10 Optimal route

The route is formed in such a way that the next stop number is the combination with the previous stop number; in this way, it is possible to know which is the point of departure and departure, in order to perform the sum of the distance and corroborate which is the optimal route. An important aspect is to solve with less time the resolution of a new neighborhood with student locations and determine the correct locations of each stop of this route, as you can see in Fig. 11.

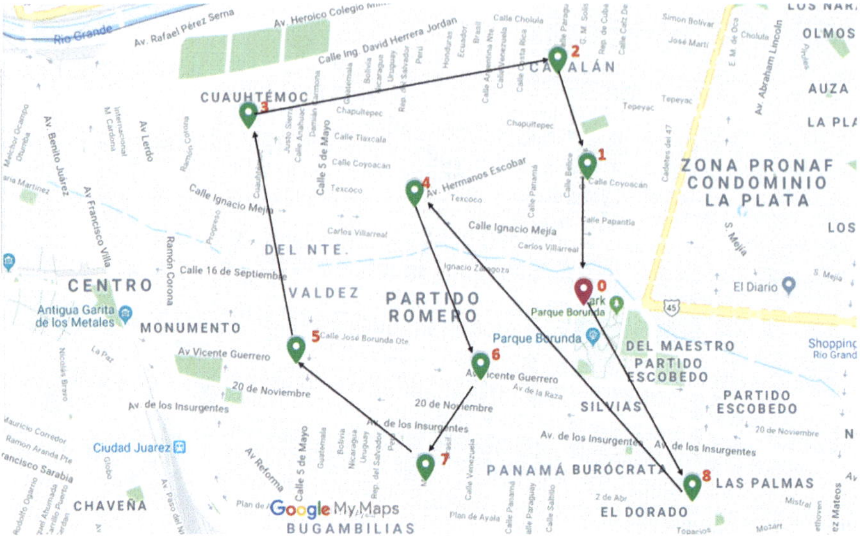

Fig. 11 Description of each group of students and their associated stops (a) and then the execution of waiting times (b)

5 Conclusions and Future Research

5.1 Regarding the Objective

In this section, we will analyze the results obtained by the library, where the route obtained will be shown, the distance between bus stops which is linear. It will show part of the code that was developed for the operation of the library and the realization of a test to confirm that the library is adaptable for any number of bus stops; it will also show the way in which the routes are saved and the total distances of them to verify that the route thrown by the library is optimal.

5.2 Testing Our Java Library Proposed

When using the library, this resulted in a first route found in the graph, which was taken as the optimum at that time; the code shown in the following figure is with which I throw this first route. In Table 3, the route with the distances of each point is shown; the minimum distance is from stop one, to stop two, with 42 meters, with the longest distance at stop three, to stop zero with 160 meters, as you can see in Fig. 12.

After some modifications to the code were made, which are shown in Fig. 13, a new route with a shorter distance was obtained; in Table 4, the improvement in the

Table 3 First route

Stop bus	Distance
[1][2]	0.00423728651379396
[2][4]	0.00826666232526575
[4][6]	0.00718901397967449
[6][7]	0.00458543400345120
[7][5]	0.00738010840028139
[5][3]	0.00911904868942050
[3][0]	0.01661762513116540
[0][8]	0.00888440712710274
[1, 8]	0.01304609811399880
Dist. Total	0.07932568428415420

Fig. 12 Code before modifying

```
min=myList[0][0];


for(int i = 0; i <myList.length-1; i++)
    {

        for(int j = 0; j < myList.length; j++ )
        {

            if(min>myList[start][j])
            {
                    min=myList[start][j];
                    nodo=j;

            }

        }

        dist= dist+min;
        min= 100.0000;

            for(int k=0; k < ruta.size();k++)
            {

                myList[nodo][(int)ruta.get(k)]=100.0000;
                myList[(int)ruta.get(k)][nodo]=100.0000;
```

route is shown, and unlike Table 3, it is possible to appreciate the decrease in total distance, and in this route, the minimum distance is from stop two, to stop one with 42 meters, with the greater distance at stop eight, to stop four with 167 meters. The minimum distances in the two tables are the same; the difference is the direction in which the points are traversed for each route.

The routes that are obtained by the library are stored in an array, in which they are saved by columns where the first number of the route is the number of the column to which it belongs. The following Table 5 shows the routes thrown by the library.

To corroborate the route that the library shows as the optimum, the distances of the other routes were added, the distances are shown in the following table where

Fig. 13 Modified code

```
for(int x=0; x<myList.length; x++)
{
   start=x;
   columna=x;
   arreglo[renglon][columna]=x;
    renglon++;
   ruta.clear();
   ruta.add(start);


   for(int i = 0; i <myList.length-1; i++)
       {

         for(int j = 0; j < myList.length; j++ )
         {

            if(min>myList[start][j])
            {
                    min=myList[start][j];
                    nodo=j;

            }

         }

         dist= dist+min;
         min= 100.0000;
```

Table 4 Optimal route

0	1	2	3	4	5	6	7	8
7.8	7.9	8.4	9.1	7.7	8.6	8.8	8.1	7.8

Table 5 Route arrangement

0	0	1	2	3	4	5	6	7	8
1	1	2	1	4	6	7	7	6	0
2	2	4	0	6	7	6	5	0	1
3	4	6	6	7	5	0	4	1	2
4	6	7	7	5	3	1	1	2	4
5	7	5	5	0	2	2	2	4	6
6	5	3	4	1	1	4	0	5	7
7	3	0	3	2	0	3	8	3	5
8	8	8	8	8	8	8	3	8	3

the distances are represented in kilometers, and the column corresponds to each one of the routes. In the Table 6, you can see that route four is effectively the shortest distance with 7.7 kilometers, and the worst route is the three with 9.1 kilometers.

Table 6 Total, distances by route

	0	1	2	3	4	5	6	7	8	9	10
0	100	0.004754	0.008854	0.016618	0.008528	0.013256	0.005438	0.009781	0.008884	0.00502	0.004781247
1	0.004754	100	0.004237	0.015505	0.007914	0.01495	0.00897	0.013555	0.013046	0.003581	0.000592864
2	0.008854	0.004237	100	0.014227	0.008267	0.016203	0.01211	0.016556	0.017278	0.00702	0.004484459
3	0.016618	0.015505	0.014227	100	0.008102	0.009119	0.014198	0.015547	0.024567	0.019018	0.016096238
4	0.008528	0.007914	0.008267	0.008102	100	0.007995	0.007189	0.010359	0.016796	0.011221	0.008497105
5	0.013256	0.01495	0.016203	0.009119	0.007995	100	0.008442	0.00738	0.018732	0.017582	0.015462907
6	0.005438	0.00897	0.01211	0.014198	0.007189	0.008442	100	0.004585	0.010616	0.010381	0.009276332
7	0.009781	0.013555	0.016556	0.015547	0.010359	0.00738	0.004585	100	0.012128	0.014798	0.01385946
8	0.008884	0.013046	0.017278	0.024567	0.016796	0.018732	0.010616	0.012128	100	0.011041	0.01281321
9	0.00502	0.003581	0.00702	0.019018	0.011221	0.017582	0.010381	0.014798	0.011041	100	0.003017095
10	0.004781	0.000593	0.004484	0.016096	0.008497	0.015463	0.009276	0.013859	0.012813	0.003017	100

Table 7 Distance with new bus stops

Stop bus	Distance
[4][6]	0.00718901397967449
[6][7]	0.00458543400345120
[5][7]	0.00738010840028139
[3][5]	0.00911904868942050
[2][3]	0.01422713077187880
[1][2]	0.00423728651379396
[0][1]	0.00475429595208644
[0][1]	0.00888440712710274
[4][8]	0.01679555968106130
Dist. Total	0.07717228511875080

5.3 *Test with More Bus Stops*

A test was carried out with more bus stops to verify that the library is fully adaptable; for the test, two stops were added randomly. Table 7 shows the distances between points and the maximum diagonal obtained with the new number of stops. The route that the library launched with these new routes is 9, 10, 1, 2, 4, 6, 7, 5, 3, 0, and 8. This new route has a total distance of 8.0 kilometers, with the minimum distance in the travel from stop ten to one with a distance of 500 meters. The following Table 8 shows the route and the distance between each point with the total distance.

When obtaining these results, it can be said that the test was satisfactory and that the library is adaptable for different numbers of stops. Besides that, the user can use the library in a simple way; when added to the project, he is using and importing the class coordinates from the library; a coordinate-type object is created in which it will serve to call the functions that are getDatos and vecinoC; the first function that has to be called is getData and creates two double-type matrices for it, because this function is the one that returns the distance matrix that will be sent to the neighborC function in which an ArrayList is used. This is because the function receives the two matrices, and the ArrayList has the optimal route that is returned by the function. In the following image, the code is shown (Fig. 14).

Table 8 Route with new distances

Bus stops	Distance
[9][10]	0.003017
[1][10]	0.000593
[1][2]	0.004237
[2][4]	0.008267
[4][6]	0.007189
[6, 7]	0.004585
[5, 7]	0.00738
[3, 5]	0.009119
[0, 3]	0.016618
[0, 8]	0.008884
[8, 9]	0.011041
Dist. Total	0.080931

Fig. 14 Call of functions

```
Coordenadas gfo = new Coordenadas();
double [][] x;
double [][] x1;
ArrayList ruta=new ArrayList();
x=gfo.getDatos();
x1=gfo.getDatos();

 ruta=gfo.vecinoC(x,x1);
```

6 Conclusions

In this research, we analyze the objective of this project, which is to develop a library programmed in Java and a minimum scheme in a relational database, which is publicly accessible for the development of an application, designed to solve the school bus routing problem; with the library, you will have a tool for the implementation and solution of the SBRP.

6.1 With Respect to the Main Objective of Our Investigation

When the library was created in the beginning, the Dijkstra algorithm was contemplated. By having the library finished with said algorithm, the expected results were not obtained, since this algorithm searches for the shortest path from an initial point to an endpoint, without going through all the points; for the user, it would be a bit more complicated to perform the optimization, since he would have to add all the adjacent points he needs with their respective distances. In this case of SBRP having nine bus stops, the route that Dijkstra throws consists of four stops with two different final destinations, one with the final destination five, resulting in stops one, two, three, and five; the other route is with the final destination six, which gives stops one, nine, seven, and six, and both routes start at stop one. When making the library with the algorithm of near neighbor, a route was obtained in which all the points are crossed, and it is easier to use by the user; since the library makes all the necessary calculations, it would only have to fill the tables in the database for the proper functioning of the library. When performing the corresponding tests and verifying the result that the library showed as the optimum, it was possible to conclude that the objective of the project was met, since there is a library that can solve the SBRP and throws the optimal path as result.

6.2 Recommendation in the Future Application of Our Research

As recommendations for this project, you can implement some tool or algorithm, to take the flow of traffic from the streets and thus stop being linear and to have a route closer to reality; another of the things that could be improved is the call of the algorithm of near neighbor, not to have to send the two matrices of distances and not to occupy that space in memory.

6.3 Future Research

Being a project that improves a route, not only can it be used for schools, but also it can be used for parcel companies, delivery companies, or transport companies. Using a mobile device associated with a more detailed framework could even determine specific days with more transported objects associated with each student, as can be seen in Fig. 15.

An aspect of utmost importance will be to adapt in time, the specifications of each individual in childhood that allows it to be competitive, helping it with mobility in a smart city, as can be seen in Fig. 16.

The route is formed in such a way that the next stop number is the combination with the previous stop number; in this way, it is possible to know which is the point of departure and departure, in order to perform the sum of the distance and corroborate which is the optimal route. An important aspect is to solve with less time the resolution of a new neighborhood with student locations and determine the correct locations of each stop of this route, as you can see in Fig. 17.

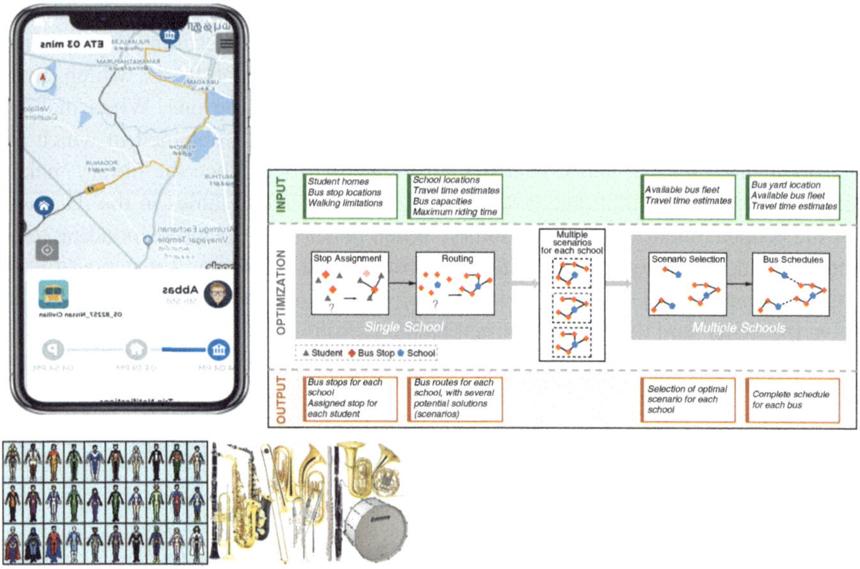

Fig. 15 Intelligent design of a mobile application based on Ubicuro Computing that would allow each student to send information associated with the transport of musical objects of a band and online specifications of each particular situation and of the dimensions and care of their respective musical instruments

Fig. 16 The OECD considers that by 2027, seven of the 12 components of a smart city will be associated with improving school performance, including student mobility

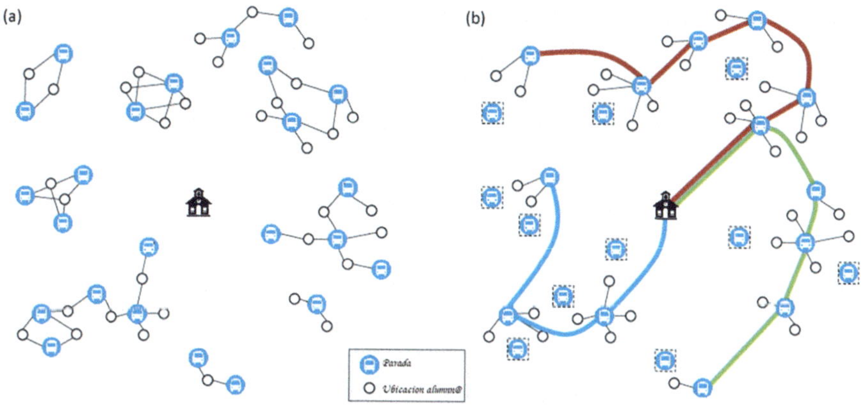

Fig. 17 Description of each group of students and their associated stops (**a**) and then the execution of waiting times (**b**)

References

1. Kang, M., Kim, S., Felan, J., Choi, H., & Cho, M. (2015). Development of a genetic algorithm for the school bus routing problem. *International Journal of Software Engineering and Its Applications, 9*(5), 107–126. Available: https://doi.org/10.14257/ijseia.2015.9.5.11.
2. Andrade, T., de Arruda Pereira, M., & Wanner, E. F. *Development of an application using a clustering algorithm for definition of collective transportation routes and times,*» DECOM – Centro Federal de Educação Tecnológica de Minas Gerais – Campus II – Brazil.
3. Jiménez, J. F., Arias-Rojas, J. S., & Montoya Torres, J. R. (2012). *Solving of school bus routing problem by ant colony optimization, Revista EIA*. [En línea]. Disponible: https://goo.gl/t2sWUT. Accedido: 17-Sep-2017.
4. Bekas, S. E. T. (2007). *Solving school bus routing problem through integer programming. The Journal of the Operational Research Society, 58(12)*, 1599–1604. Palgrave Macmillan Journals.
5. Bögl, M., Doerner, K. F., & Parragh, S. N. (2015). *The school bus routing and scheduling problem with transfers*. [En línea]. Disponible: https://goo.gl/rQ9tcJ. Accedido: 23-Sep-2017.
6. Yingit, T., & Unsal, O. (2016). *Using the ant colony algorithm for real-time. The International Arab Journal of Information Technology, 13*(5). [En línea]. Disponible: https://goo.gl/z2xfvM. Accedido: 23-Sep-2017.
7. De Souza Lima, F. M. (2015). *A mixed load rural school bus routing problem whit heterogeneous fleet: A study for the brazilian problem*, tesis, Universidade Federal de Minas Gerais. [En línea]. Disponible en: https://goo.gl/DGdiJj. Accedido: 27-Sep-2017.
8. de Guevara, J. M. L. *Fundamentos de programación en Java*, Editorial EME. [En línea]. Disponible: https://goo.gl/tWUr2D. Accedido: 26-oct-2017.
9. Groussard, T.. (2012). *JAVA 7: Los fundamentos del lenguaje Java*, Ediciones ENI. [En línea]. Disponible: https://goo.gl/ak7yfv. Accedido: 26-Oct-2017.
10. Quintana, G., Marqués, M., Aliaga, J. I., & Aramburu, M. J. (2008). *Aprende SQL*, Publicacions de la Universitat Jaume I. [En línea]. Disponible: https://www.e-buc.com/portades/9788480217729_L33_23.pdf. Accedido: 26-Oct-2017.

11. Wilton, P., & Colby, J. (2005). *Beginning SQL* (p. 12). Indianapolis: Wiley.
12. Mehlhorn, K., Orlin, J., & Tarjan, R. (1987). *Faster algorithms for the shortest path problem*. Technical report CS-TR-154-88, Department of Computer Science, Princeton University. [En línea]. Disponible: ftp://ftp.cs.princeton.edu/reports/1988/154.pdf. Accedido: 26-Oct-2017.
13. Park, J., & Kim, B. I. (2010). The school bus routing problem: A review. *European Journal of Operational Research, 202*(2), 311–319. [En línea]. Disponible: https://bit.ly/2PrrQae. Accedido: 29-Oct-2017.
14. Gutin, G., Yeo, A., & Zverovich, A. (2002). Traveling salesman should not be greedy: Domination analysis of greedy-type heuristics for the TSP. *Discrete Applied Mathematics, 117*(1–3), 81–86. [En línea]. Disponible: https://bit.ly/2C1iRno. Accedido: 02-Nov-2018.
15. Cunquero, R. M. (2003). *Algoritmos heurísticos en optimización combinatoria*. Universidad de Valencia, Facultad de Ciencias Matemáticas. [En línea]. Disponible: https://bit.ly/2TxPHTT. Accedido: 14-Dic-2018.
16. Torrubia, G., & Terrazas, V. (2012). *Algoritmo de Dijkstra. Un tutorial interactivo*. VII Jornadas de Enseñanza Universitaria de la Informática (JENUI 2001). [En línea]. Disponible: https://bit.ly/2PqhDrI. Accedido: 14-Dic-2018.
17. Sniedovich, M. (2006). Dijkstra's algorithm revisited: the dynamic programming connexion. *Control and Cybernetics, 35*((3), 599–620. [En línea]. Disponible: https://bit.ly/2JbVIEb. Accedido: 14-Abr-2019.

An Efficient Hybrid Evolutionary Algorithm for the Smart Vehicle Routing Problem

Hajer Ben-Romdhane and Saoussen Krichen

1 Introduction

The evolution in the concept of city over the decades and the rise of innovation on the urban landscape yielded to an urgent necessity of thinking about a best or a "smart" management of the city for improving citizens' well-being. The concept of smart cities has then emerged to describe sustainable development and life quality offered by advanced technologies in all domains, including economy, transportation, health care, waste management, and energy consumption. Given its fundamental importance, smart mobility is one of the most fast-growing areas using various information and communication technologies (ICTs) actively. A well-known example of smart mobility is the intelligent transport system (ITS) [24, 34] which helps optimizing people and goods mobility while considering comfort, safety, road traffic, and cost savings. These systems operates by collecting information from different sources, e.g., Global Positioning System (GPS), Global Navigation Satellite System (GNSS), and Cooperative Intelligent Transport System (C-ITS), processing the collected data using a decision-making tool, and communicating solutions to users. These decision support systems are deployed in several domains as transportation (traffic light control and vehicle routing) [24], military (predation missions using drones) [3], and supply chain management [27].

The current work is especially concerned with real-time routing of vehicles in main cities. This problem is a prominent and very challenging task, especially when considering roads conditions, traffic jam, and dynamic demands from customers. So far, a lot of efforts have been expended in designing ITS and proposing efficient decision-making tools for vehicle routing. But, very few works have addressed

H. Ben-Romdhane (✉) · S. Krichen
Larodec Laboratory, Institut Supérieur de Gestion de Tunis, Université de Tunis, Le Bardo, Tunisia

M. M. Banat, S. Paiva (eds.), *Smart Technologies for Smart Cities*, EAI/Springer Innovations in Communication and Computing,
https://doi.org/10.1007/978-3-030-39986-3_10

mobility of vehicles in both a green and dynamic environment. Indeed, in smart cities, vehicle mobilities have to consider, besides cost savings, an environment-friendly solutions, or what is known as "green mobility." This requires active mobilization of efforts toward the mitigation of climate change and more precisely in reducing energy consumption main cause of air pollution. Our proposal here is a model that aims to improve drivers' experience in modern cities while respecting the environment. Accordingly, the two main contributions of this study are (1) to model the problem mathematically and (2) to propose an efficient solution approach to solve it.

The afore-discussed problem is modeled as a dynamic energy-minimizing vehicle routing problem with time windows (D-EVRPW). More formally, it is defined as follows: a number of vehicles serving a set of dispersed customers while minimizing the total energy consumption, respecting the vehicles capacities, and the customers service time windows. By means of the shared and collected data, changes in the planned orders (e.g., canceled order, new customer requests) and in the environment conditions (e.g., closed road, traffic jam) are revealed overtime and are to be considered in order to continuously adapt the routes to the new conditions and assist the driver for choosing the best road.

To solve D-EVRPW, a new hybrid evolutionary algorithm (EA) is proposed. As exploration and exploitation (E&E) are the two antagonistic search modes that sway EAs efficiency [31], a good balance between E&E would be the key forward a successful search engine. Moreover, EAs are often challenged by convergence to a local optimum which may break the adaptation after environmental changes. In such cases, neither exploration nor exploitation can be pursued exclusively to easily adapt to the new environment. The solution would be to create an adaptive balance between exploration and exploitation to better steer the search direction; e.g., more exploration is needed after changes, while exploitation is required to reinforce the search in specific areas. We propose here a new scheme for adaptively balancing exploration and exploitation in EAs: based on the results of a statistical Pre-Post analysis of the population, the next search mode can be decided. That is, if the current search mode failed to improve the population, the algorithm shifts from a phase of exploration to a phase of exploitation and inversely. In order to validate the proposed approach, a series of experiments is conducted to compare it to several competing approaches and in terms of two different objectives: minimization of the traveled distance and minimization of the energy consumption. The experimental investigation show that the proposed approach excels versus them in most cases.

The rest of this article is organized as follows. Section 2 briefly reviews the literature related to studied problem. In Sect. 3, the proposed model is described and its mathematical formulation is presented. Section 4 presents the proposed approach and outlines its pseudocode. The experimental results and their analysis are provided in Sect. 5. Section 6 is devoted to the concluding remarks and perspectives.

2 Related Literature

To better ground the study proposals, this section presents a related literature review. First, a summary of the most important variants and extensions of the vehicle routing problem (VRP) is provided. Second, a brief but consistent review of the existing evolutionary approaches for dynamic optimization problems is presented.

2.1 *A Review on the Vehicle Routing Problem*

The VRP is one of the most studied combinatorial optimization problems due to its ability to model complex real-world situations [36]. It has known many extensions and variations over the years depending on the nature of the tackled problems. Figure 1 presents a new taxonomy of VRP variants according to four aspects: additional specific constraints, number of objectives, type or time of service, and environment-friendly problems.

From the first category, the capacitated VRP (CVRP) is one of the primitive versions of the problem that prohibit the violation of the vehicles capacity [21]. Besides, the distance-constrained VRP (DCVRP) have been considered an additional constraint which sets a maximum allowed distance for the routes [19].

The second aspect includes the number of tackled objectives. Basically, the original version of VRP consists of minimizing a single objective (e.g., cost, distance, or number of vehicles) [21, 22]. Multi-objective VRP, however, considers two to three objectives to optimize concurrently as minimizing the waiting time of vehicles and maximizing customers' preferences [15] or minimizing both the routes cost and risk [9]. Moreover, many-objective VRPs consider more than three objectives [25, 32].

Another category of problems have addressed other types of services other than the delivery of goods. The VRP with pickup and delivery (VRPPD) [10, 22] allows transporting goods from the depot to customers but also from customers to the depot. Besides, some studies have focused on the VRP with backhauls (VRPB) in which customers are allowed to return purchased commodities [13]. A more complex problem investigated the delivery of demands to customers within specific time windows (VRPTW), which specifies the interval of time in which a customer should be served [22, 35].

More recently, a new trend of works has started to develop environment-friendly transportation solutions. Thus, many studies have proposed models for optimizing the mobility of vehicles carrying dangerous materials. As the transportation of such products may be threatening to the environment (e.g., flammable substances), the hazardous material VRP (HVRPs) aim to minimize the risk associated with their transportation [9, 41]. Green VRPs are also considered to be eco-friendly solutions with different objectives: reducing CO2 emissions [8, 23] and minimizing energy consumption [18].

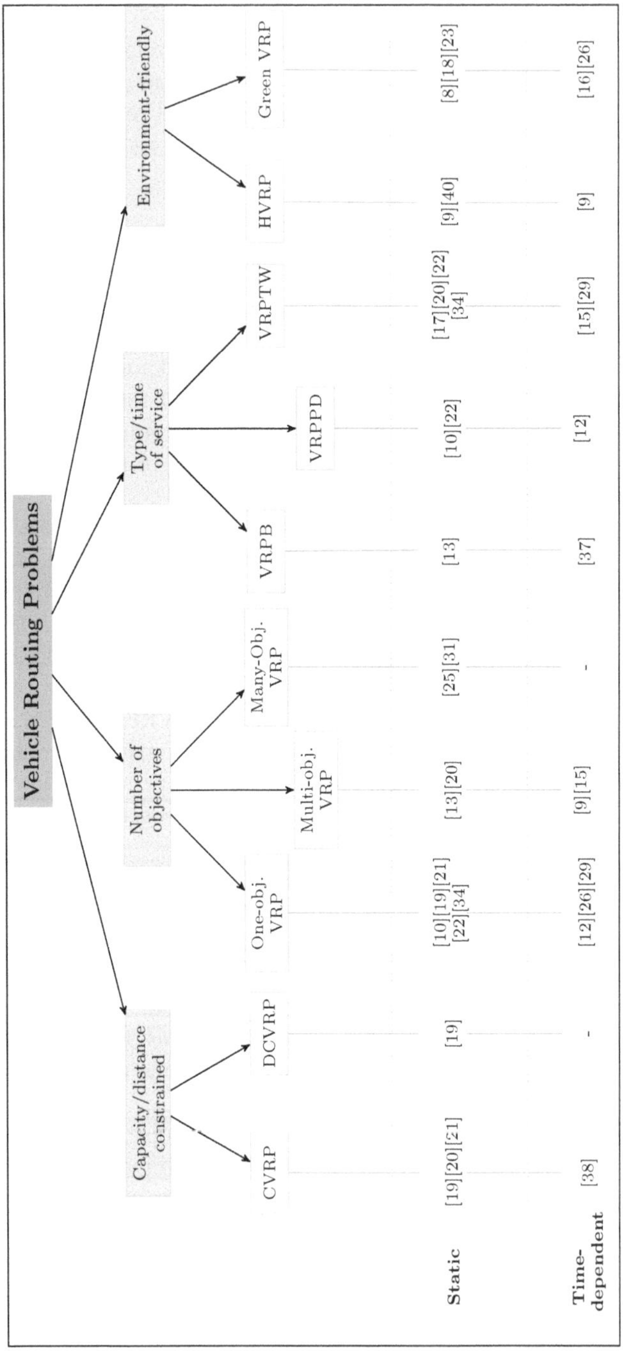

Fig. 1 Taxonomy of VRP variants

In recent years, and with the development of ICTs, real-time routing of vehicles becomes affordable. Many variants of the VRP have been studied under dynamic environments where new customers' requests may be revealed (e.g., new customers, variation in the ordered quantity of planned customers), or road conditions may vary (e.g., closed road, traffic jam) after the vehicle starts to serve customers [30]. Therefore, the planned route has to be rearranged to take into consideration the new environment conditions. Most of the studied problems have focused on real-time mobility using road traffic prediction for traffic lights control [6, 40]. And very few works have been concerned with real-time mobility in eco-friendly environment. In [16] and [26], dynamic green VRPs are studied with the objective of minimizing the CO2 emissions, and heuristic methods based on the genetic algorithm and the simulated annealing approaches were proposed to solve the problem. The transportation of hazardous materials while considering both static and dynamic travel times have been addressed in [9], and several heuristic procedures have been tested to solve it.

Despite the efforts expended toward the investigation of real-time routing, the existing studies have focused on one or two aspects of the problem. A more interesting variant of the problem in the context of "smart mobility" would be the one that considers real-time and eco-friendly routing of vehicles while taking into account customers' satisfaction. To this end, we study in this work a dynamic VRP with the objective of minimizing the energy consumption while serving the customers in their preferred time interval (time windows).

2.2 *Evolutionary Algorithms for DOPs*

EAs are among the most widespread and successful optimization approaches that have addressed combinatorial optimization problems. However, the major concerns with EAs in dynamic environments are how to avoid convergence and continuously adapt to the changing environments.

Over the last two decades, extensive efforts have been deployed to improve EAs abilities toward fast adaptation and tackling premature convergence. Most common approaches are based on diversity control mechanisms. Approaches belonging to this category either introduce diversity after environmental changes [7] (e.g., random re-initialization of the population, a mutation step with high rate), or maintain diversity during the run [28] (e.g., random immigrants).

A more recent trend of approaches has addressed the convergence problem by controlling the exploration-exploitation trade-off. In [37], the authors proposed an approach that switches between phases of exploration and exploitation by means of the genetic operators: exploration is performed using the mutation operator, while the selection and crossover operators are used for exploitation. Another approach consists of using multi-population to balance exploration and exploitation [5]. Most of multi-population approaches allocate a second (auxiliary) population as a memory of past solutions or as a diversity reservoir. In [4], two subpopulations are used for each search mode while switching between the two modes iteratively.

While the last type of approaches showed a good performance, it cannot be deemed to be adaptive as the control of the search mode is deterministic. Alternatively, the proposal of this paper is self-adaptive in the sense that the appropriate search mode is dynamically determined based on the feedback of the algorithm itself.

3 Problem Description

The dynamic energy-minimizing vehicle routing problem with time windows (D-EVRPW) models a real-time routing of a number of homogeneous vehicles serving a set of customers within predefined time windows while minimizing the energy consumption. When considering the dynamic aspect of the environment, the problem becomes time-dependent, and the planned routes should be adapted to convey to the new environment conditions (e.g., closed road, traffic jam) and customers' requests (e.g., canceled order, new customer requests).

More formally, n dispersed customers, each characterized by spatial coordinates (x_i, y_i), a demand q_i (quantity of goods), and a time window (earliest and latest time of service) are to be served by a set ϑ of homogeneous vehicles of a limited capacity Q, starting and ending at the same depot (with coordinates (0,0)). The objective is to minimize the energy consumption of the used fleet of vehicles.

In the literature, various estimates of the energy function have been proposed. As the weight of the transport vehicles is directly related to their energy consumption, the minimization of the weight of the vehicles will reduce the amount of the energy they use. Therefore, and for the sake of simplicity and computation easiness, the energy formula proposed in [18] is adopted. It is given by:

$$Energy(i, j) = [Tare + \text{Load of the vehicle over}(i, j)] * \text{distance}(i, j) \quad (1)$$

where the energy consumption of a vehicle between customers i and j is computed as the weight of the empty vehicle plus the load of the vehicle between i and j and weighted by the traveled distance. The load of the vehicle is given by the total amount of commodities to be delivered to the customers starting from i, while the distance between two points is computed as the Euclidian distance:

$$d_{ij} = \sqrt{(x_i - x_j)^2 + (y_i - y_j)^2} \quad (2)$$

where (x_i, y_i) and (x_j, y_j) are, respectively, the coordinates of customers i and j in a two-dimensional Cartesian coordinate system.

In the following table, we enumerate the relevant notations that are being used next in the mathematical formulation of the problem (Table 1).

Table 1 List of notations

Parameters	Description
n	Set of customers
ϑ	Set of vehicles
ν	Set of vertices
Q	Capacity of a vehicle
Q_0	Weight of the empty vehicle
q_i	Demand of customer i, where $i \in \{1, .., n\}$
d_{ij}	Distance from node i to j, where $i \neq j$
r_{ij}	Travel time from customer i to customer j
s_{ik}	Time vehicle k starts to service customer i
a_i	Earliest time for serving customer i
b_i	Latest time for serving customer i
x_{ijk}	Takes 1 if vehicle k travels from i to j, and 0 otherwise
y_{ij}	Weight of a vehicle between i and j, and 0 otherwise

The mathematical model of the D-EVRPW, inspired by the VRP with time windows formulation proposed in [17], is stated as:

$$Min \sum_{i\in\nu}\sum_{j\in\nu} d_{ij} y_{ij} \tag{3a}$$

$$S.T$$

$$\sum_{k\in\vartheta}\sum_{j\in\nu} x_{ijk} = 1, \quad \forall i \in n \tag{3b}$$

$$\sum_{i\in n} q_i \sum_{j\in\nu} x_{ijk} \leq Q, \quad \forall k \in \vartheta \tag{3c}$$

$$\sum_{j\in\nu} x_{0jk} = 1, \quad \forall k \in \vartheta \tag{3d}$$

$$\sum_{i\in\nu} x_{ihk} - \sum_{j\in\nu} x_{hjk} = 0, \quad \forall h \in n, \forall k \in \vartheta \tag{3e}$$

$$\sum_{i\in\nu} x_{i,n+1,k} = 1, \quad \forall k \in \vartheta \tag{3f}$$

$$x_{ijk}(s_{ik} + r_{ij} + s_{jk}) \leq 0, \quad \forall i, j \in \nu, \forall k \in \vartheta \tag{3g}$$

$$a_i \leq s_{ik} \leq b_i, \quad \forall i \in \nu, \quad \forall k \in \vartheta \tag{3h}$$

$$y_{0i} = Q_0 x_{0ik} \tag{3i}$$

$$y_{ij} < (Q + Q_0 + q_j) x_{ijk}, \quad (i, j) \in A \tag{3j}$$

$$y_{ij} \geq (Q_0 + q_j)x_{ijk}, \quad (i, j) \in A \tag{3k}$$

$$x_{ijk} \in \{0, 1\}, \quad \forall i, j \in \nu, \quad \forall k \in \vartheta \tag{3l}$$

where:

- **Equation (3a)** refers to the objective function which aims to minimize the total amount of energy used by the set of vehicles.
- **Constraints (3b)** indicate that each customer is visited exactly once.
- **Constraints (3c)** ensure that the capacity of the vehicles is not exceeded.
- **Constraints (3d), (3e), (3f)** indicate respectively that each vehicle must depart from the depot, that once a customer is served the vehicle should leave for another customer, and that all vehicles should return to the depot.
- **Constraints (3g)** states the arrival time of a vehicle k departing from customer i to j at $(s_{ik} + r_{ij})$.
- **Constraints (3h)** guarantee that a customer i has to be served within a specific time window.
- **Constraints (3i)** initialize the flow on the first arc of each route.
- **Constraints (3j) and (3k)** set the upper and lower bounds for the flow of any road.
- **Constraints (3l)** ensures the integrality of the decision variables.

As we are dealing with a dynamic environment, the decisions variables and several parameters are subject to variation over time. Therefore, depending on the complexity of the problem, some (or all) of the following parameters are to be considered time-dependent:

- the total number of customers (requests from new customers may be revealed over time, and other demands may be canceled)
- the total number of available vehicles (may decrease because of a vehicle breakdown)
- the demand of a given customer (can increase or decrease)
- the load of the vehicle between i and i (depending of the total quantity of demands which is also varying over time)
- the total number of vertices (e.g., because of a road closure, poor road quality)
- time windows of customers (e.g., may vary in holidays or in summer)

4 PPTEA: An Efficient Hybrid EA for the D-GVRP

The success of an EA relies heavily on its ability to achieve an effective balance between exploration and exploitation. Numerous studies have addressed this issue in the literature and proposed different approaches to solve it [1, 4]. However, coming up with the ideal ratio between two competing search activities is not a trivial task. This is due to the difficulties involved in the measurement and control of such a trade-off. Several questions are worth raising in this regard: "How to determine the activity to engage at a given point in time?" "How long should an activity last?" "How to measure its effectiveness?" and "When to switch to the other search mode?"

In order to address these questions, we introduce PPTEA: a new EA with population Pre-Post testing. In PPTEA, the population is split into two explicit subpopulations to stress each search behavior (exploration and exploitation) separately with sparse interactions. The algorithm switches between the two search modes in an adaptive manner and according to the following principle: after the accomplishment of a search activity (either exploration or exploitation), the population undergoes a Pre-Post testing [11] (using the Wilcoxon signed-rank test) to determine significant improvement or stagnation in the fitness landscape. If it is still improving, the algorithm continues with the current search mode; otherwise it switches toward the other search mode.

Since population-based methods are heavy on diversification and trajectory-based methods are powerful on intensification, the exploration activity will be handled by a genetic algorithm (GA), while the exploitation activity is assigned to a local search (LS). The GA considered here is just a basic steady-state GA in which the three genetic operators (selection, crossover, and mutation) are used to create a new offspring that replaces the worst of the current population. As for the LS, a Hill-climbing algorithm is used: it creates new solutions by swapping the positions of two customers in the selected solution and then introduces the best of them in the current population. Thus, the two-subpopulations are evolved independently during exploration and exploitation activities, and the migration of individuals is allowed at the end of each search activity. The migration policy used here is the "best-worst" which consists of selecting the best individual of one population to replace the worst of the recipient subpopulation.

Besides, and in order to better tackle the dynamics of the environment, increasing the diversity of the population is needed after environmental changes. Therefore, the current population goes through a readaptation (to take into consideration the new requests), and a small rate of individuals are replaced by randomly generated ones. The main steps of PPTEA are summarized in the schema of Fig. 2.

5 Experimental Study

In this section, the experimental results are presented and analyzed. In order to evaluate the relevance of the proposed approach, we compare PPTEA to several algorithms according to the considered objective (minimization of the energy consumption) and also the minimization of the total distance, separately. Our choice is motivated by the fact that the best-known results of the used benchmarks exist only for the minimization distance objective. Therefore, the obtained results are compared to the results of a general variable neighborhood search (GVNS) [2], as well as the results of a simple genetic algorithm and an evolutionary local search (a hill-climbing) so as to assess the "components" of PPTEA separately.

```
Initialize sub-populations P_1 and P_2 at random
mode ← "explore"
while Stopping criterion not met do
   if Environmental change is detected then
      for each p in {1,2} do
         Adapt(P_p(g), r_rand, n)
      end for
      mode ← "explore"
   end if
   if mode=="explore" then
      Perform an exploration phase /* Steady state GA */
   else
      Perform an exploitation phase /* Hill-climbing LS */
   end if
   if ( PrePost-Test(P(g),P(g − 1))=="No improvement") then
      SwitchSearch(mode) /* From exploration to exploitation, and inversely */
      InterPop-Immigration(P_p(g), P_{3−p}(g))
   end if
   g ← g + 1
end while
```

Fig. 2 Pseudocode of PPTEA

In the following, there will first be a description of the problem instances used in this work. Then, the experimental settings are presented. And finally, the computational results will be reported and analyzed.

5.1 Problem Instances

The computational experiments were carried out on the well-known test instances of VRPTW by M. Solomon [33]. It consists of 56 different problem instances, with 100 customers, each of which is associated with the following data: customer number, Cartesian coordinates (integer values ranging from 1 to 100), the ordered quantity, earliest time, due time, and service time. The geographical data used here belongs to the following categories: R-type (uniformly distributed customers), C-type (clustered customers) and RC-type (a mix of R and C types) with different scheduling horizons: R101, C101, and RC101 indicate problems with short scheduling horizon, while R201, C201 and RC201 contain data with long scheduling horizon.

As these data instances are static, the following steps are used to transform them into dynamic instances (as proposed in [14]). First, the set of n customers is split into two subsets: customers' requests that are known beforehand n_s and dynamic customers' requests that are revealed over time n_d. The size of each subset is fixed based on the following formula:

$$dod = \frac{n_d}{n_s + n_d} \tag{4}$$

Equation (4) defines the degree of dynamism dod which means the severity of the change in environment (measured by the number of new requests revealed online with regards to the total number of requests) [30]. Once the initial set of customers is identified, initial routes can be optimized to serve the known customers. Then, when new requests are revealed, the routes are rearranged in such a way to consider the remaining customers (known demands but not yet served) as well as the new demands.

5.2 Design of the Experimentations

In the experiments, 30 independent runs are performed for each data instance. The stopping criteria is the maximum number of evaluations. The parameters settings of the implemented algorithms are shown in Table 2.

5.3 Experimental Results

We present and analyze, in this section, the performance results of the considered algorithms. The numerical results related to the two considered objectives (total distance minimization and energy consumption minimization) are provided, respectively, in Tables 3 and 4. Bold-faced values indicate the best performing algorithm. The last row of the table shows the results of the normalized score (NS), which normalizes the results obtained by each algorithm over the number of problem instances by scaling them between 0 and 1 (the higher is NS, the better is the algorithm). As we are dealing with a minimization problem, the NS is given by:

$$NS(A) = \frac{1}{T} \sum_{j=1}^{T} \frac{| \overline{A}(j) - \overline{Worst}(A, j) |}{| \overline{Best}(j) - \overline{Worst}(j) |} \tag{5}$$

where T is the total number of configurations (in our case $T = 30$) and $\overline{A}(j)$ is the fitness value of algorithm A in the jth configuration, while $\overline{Best}(j)$ and $\overline{Worst}(j)$

Table 2 Parameters settings

Total number of evaluations	1000
Population size	100 individuals
Selection	Tournament selection
Crossover	Order crossover ($p_c = 1$)
Mutation	Swap mutation ($p_m = 0.1$)
Replacement operator	The worst in the current population
Replacement rate	$r_{rand} = 0.3$

Table 3 Comparison of the obtained results in terms of total distance

			Total distance			
Problem instance	*Q*	*dod*	GA	LS	GVNS [2]	PPTEA
R101	200	0.1	1255.91	1253.12	1250.16	**1245.16**
		0.3	1280.32	1286.36	1256.04	**1251.13**
		0.5	1285.57	1281.53	**1267.47**	1271.74
		0.7	1312.53	1297.34	1267.78	**1263.59**
		0.9	1277.01	1267.14	1250.38	**1250.01**
C101	200	0.1	**894.55**	897.77	898.30	897.15
		0.3	937.79	962.08	949.95	**932.59**
		0.5	1006.23	993.83	992.97	**989.74**
		0.7	1035.37	1022.68	1009.47	**1003.52**
		0.9	1035.25	1031.82	963.33	**954.96**
RC101	200	0.1	1450.04	1436.23	1417.07	**1410.72**
		0.3	1498.96	1495.23	1471.00	**1445.13**
		0.5	1499.02	1501.72	**1484.01**	1489.69
		0.7	1497.23	1497.10	1489.28	**1479.42**
		0.9	1525.41	1520.04	1470.45	**1465.66**
R201	200	0.1	1002.31	992.0	1000.00	**983.27**
		0.3	999.11	995.78	1035.61	**975.05**
		0.5	1075.07	1056.52	1071.83	**1009.31**
		0.7	1077.09	1049.04	1078.03	**1021.34**
		0.9	1095.08	1071.82	1086.78	**1041.27**
C201	1000	0.1	666.17	676.71	668.99	**627.72**
		0.3	689.38	686.40	672.95	**636.01**
		0.5	659.11	630.18	**623.10**	624.52
		0.7	691.42	686.72	**672.95**	673.10
		0.9	674.19	681.85	668.99	**665.23**
RC201	700	0.1	1120.12	1103.30	1153.43	**1102.50**
		0.3	1166.17	1159.45	1172.33	**1133.54**
		0.5	1198.04	1193.15	1200.26	**1169.25**
		0.7	1221.48	1198.16	1234.36	**1159.12**
		0.9	1237.92	1217.78	1275.93	**1189.36**
Normalized Score			0.16	0.31	0.48	**0.96**

represent, respectively, the best and the worst fitness values obtained by all the considered algorithms.

We start by evaluating the performance of the proposed approach PPTEA according to the first objective: distance minimization. The results in Table 3 show that PPTEA outperforms the competing approaches in almost all cases. NS measure indicates that PPTEA is substantially better than the rest of approaches and that GVNS comes second in terms of performance. Besides, it can be argued that the way in which PPTEA managed the balance between exploration and exploitation

Table 4 Comparison of the algorithms performance according to the energy consumption minimization objective

Problem instance	Q	*dod*	Energy consumption GA	LS	PPTEA
R101	200	0.1	142294.60	141643.13	**141076.62**
		0.3	130976.73	**127990.59**	128492.89
		0.5	234487.96	232286.52	**231965.37**
		0.7	303456.93	293110.73	**292142.01**
		0.9	352452.00	345104.88	**345002.76**
C101	200	0.1	**231151.72**	232120.72	231823.56
		0.3	308251.57	312248.56	**306542.33**
		0.5	235457.82	232354.98	**231599.16**
		0.7	172699.72	168379.59	**167387.14**
		0.9	228272.62	212414.26	**210568.68**
RC101	200	0.1	281887.77	275478.41	**274243.96**
		0.3	309835.03	304055.70	**298708.37**
		0.5	236095.65	**233731.57**	234626.17
		0.7	297799.04	296217.79	**294256.63**
		0.9	285556.75	275268.24	**274371.55**
R201	200	0.1	238148.85	237600.00	**233624.95**
		0.3	158259.02	164039.04	**154447.92**
		0.5	204167.86	203433.33	**191567.03**
		0.7	273365.44	273604.01	**259216.09**
		0.9	285815.88	**271771.47**	283649.58
C201	1000	0.1	211842.06	212738.82	**199614.96**
		0.3	94100.37	91857.67	**86815.36**
		0.5	114553.31	108541.57	**108294.78**
		0.7	95415.96	92887.80	**92867.12**
		0.9	98633.99	97873.24	**97323.14**
RC201	700	0.1	237465.44	244527.16	**233730.11**
		0.3	193584.22	194606.78	**188167.64**
		0.5	431294.40	432093.64	**420930.25**
		0.7	215835.52	218111.41	**204816.50**
		0.9	245616.52	**228951.81**	238299.61
Normalized Score			0.13	0.46	**0.95**

through GA and LS has succeeded to improve the results obtained by each of these algorithms separately. The superiority in the performance of PPTEA over GVNS is mainly by dint of its ability to maintain a balanced search: while GVNS focuses mainly on exploitation of the neighborhood of the obtained results, PPTEA switches the search mode from exploitation to exploration adaptively all along the search process. This creates a continuous trade-off between the two search modes and prevents the population to be trapped in local optima.

Table 4 shows a comparison of GA, LS, and PPTEA in terms of energy minimization. It is apparent from the table that, also with this objective, PPTEA performs much better than its competitors. According to the NS measure, PPTEA is largely the best, and LS outperforms GA in overall.

In conclusion, it can be deemed that PPTEA is a very promising algorithm as it has shown very good results in terms of two different objectives and over a large set of instances. Besides, PPTEA has improved the best-known solutions previously obtained by GVNS by a significant margin and outperformed the results obtained by its two components GA and LS. A number of factors play a role in explaining these findings. First, balancing exploration and exploitation in PPTEA by means of the Pre-Post test helped to better steer the search direction to the needed mode and prevented convergence to local optima. Second, maintaining a diversified population is a significant contributory factor to the adaptation to the new environment conditions in dynamic optimization problems. And finally, the hybridization of GA and LS in PPTEA has proven that such combinations are needed for a really accurate and efficient algorithm.

6 Conclusion

In this study, a new VRP model for smart mobility is introduced and a new hybrid EA is proposed to solve it. Unlike most previous studies, our objective was to minimize the vehicles energy consumption rather than the total traveled distance. Besides, a new hybrid EA is proposed to solve the problem with an explicit balance between two approaches (GA and LS) to better steer the search direction especially that we are dealing with a dynamic environment. Experiments carried out on well-known benchmarks showed that our proposal has improved existing results and that it is a very promising for dynamic optimization problems. The proposed solution is intended to be part of an intelligent support system embedded in cars, to provide the driver with a real-time routing plan for a more efficient use of energy and also time saving.

As perspectives, it seems interesting to investigate the D-EVRPW with two objectives concurrently: minimize the total distance and minimize the energy consumption, for a better compromise between cost savings and environment protection. Besides, we would like also to evaluate the proposed model over real-data instances and to test the proposed approach on different dynamic optimization problem to be able to better evaluate its performance.

References

1. Alba, E., Ben-Romdhane, H., Krichen, S., & Sarasola, B. (2013). Bipop: A new algorithm with explicit exploration/exploitation control for dynamic optimization problems. In: S. Yang & X. Yao (Eds.), *Evolutionary computation for dynamic optimization problems* (pp. 171–191). Berlin/Heidelberg: Springer.

2. de Armas, J., & Melián-Batista, B. (2015). Variable neighborhood search for a dynamic rich vehicle routing problem with time windows. *Computers & Industrial Engineering, 85*, 120–131.
3. Ben-Romdhane, H., Alba, E., & Krichen, S. (2016). Towards a dynamic modeling of the predator prey problem. *Applied Intelligence, 44*(4), 755–770.
4. Ben-Romdhane, H., Krichen, S., & Alba, E. (2017). A bi-population based scheme for an explicit exploration/exploitation trade-off in dynamic environments. *Journal of Experimental and Theoretical Artificial Intelligence, 29*(3), 453–479.
5. Branke, J., Kaussler, T., Schmidt, C., & Schmeck, H. (2000). A multi-population approach to dynamic optimization problems. In: *Adaptive computing in design and manufacturing* (pp. 299–308). London/New York: Springer.
6. Chakraborty, P. S., Sinha, P. R., & Tiwari, A. (2015). Real time optimized traffic management algorithm for intelligent transportation systems. In *2015 IEEE International Conference on Computational Intelligence Communication Technology* (pp. 744–749).
7. Cobb, H. G. (1990). An investigation into the use of hypermutation as an adaptive operator in genetic algorithms having continuous, time-dependent nonstationary environments. Technical Report AIC-90-001, Naval Research Laboratory. citeseer.ist.psu.edu/cobb90investigation.html.
8. de Oliveira da Costa, P. R., Mauceri, S., Carroll, P., & Pallonetto, F. (2018). A genetic algorithm for a green vehicle routing problem. *Electronic Notes in Discrete Mathematics, 64*, 65–74.
9. Cuneo, V., Nigro, M., Carrese, S., Ardito, C. F., & Corman, F. (2018). Risk based, multi objective vehicle routing problem for hazardous materials: A test case in downstream fuel logistics. *Transportation Research Procedia, 30*, 43–52. EURO Mini Conference on Advances in Freight Transportation and Logistics.
10. Desaulniers, G., Desrosiers, J., Erdmann, A., Solomon, M. M., & Soumis, F. (2001). The vehicle routing problem. In *VRP with pickup and delivery* (pp. 225–242). Philadelphia: Society for Industrial and Applied Mathematics.
11. Dimitrov, D., & Rumrill, P. (2003). Pretest-posttest designs and measurement of change. *Work, 20*(2), 159–165.
12. Fabri, A., & Recht, P. (2006). On dynamic pickup and delivery vehicle routing with several time windows and waiting times. *Transportation Research Part B: Methodological 40*(4), 335–350.
13. Garcia-Najera, A. (2012). The vehicle routing problem with backhauls: A multi-objective evolutionary approach. In J. K., Hao & M. Middendorf (Eds.), *Evolutionary computation in combinatorial optimization* (pp. 255–266). Berlin/Heidelberg: Springer.
14. Gendreau, M., Guertin, F., Potvin, J. Y., & Taillard, E. (1999). Parallel tabu search for real-time vehicle routing and dispatching. *Transportation Science, 33*(4), 381–390.
15. Ghannadpour, S. F., Noori, S., Tavakkoli-Moghaddam, R., & Ghoseiri, K. (2014). A multi-objective dynamic vehicle routing problem with fuzzy time windows: Model, solution and application. *Applied Soft Computing, 14*, 504–527.
16. Hooshmand, F., & MirHassani, S. A. (2019). Time dependent green VRP with alternative fuel powered vehicles. *Energy Systems, 10*(3), 721–756.
17. Kallehauge, B., Larsen, J., Madsen, O. B., & Solomon, M. M. (2005). *Vehicle routing problem with time windows* (pp. 67–98). Boston: Springer.
18. Kara, İ., Kara, B. Y., & Yetis, M. K. (2007). Energy minimizing vehicle routing problem. In: A. Dress, Y. Xu, & B. Zhu (Eds.), *Combinatorial optimization and applications* (pp. 62–71). Berlin/Heidelberg: Springer.
19. Kek, A. G., Cheu, R. L., & Meng, Q. (2008). Distance-constrained capacitated vehicle routing problems with flexible assignment of start and end depots. *Mathematical and Computer Modelling, 47*(1), 140–152.
20. Konstantinidis, A., Pericleous, S., & Charalambous, C. (2014). Adaptive evolutionary algorithm for a multi-objective VRP. *International Journal on Engineering Intelligent Systems, 22*, 145 162.

21. Letchford, A. N., & Salazar-González, J. J. (2019). The capacitated vehicle routing problem: Stronger bounds in pseudo-polynomial time. *European Journal of Operational Research, 272*(1), 24–31.
22. Li, H., & Lim, A. (2001). A metaheuristic for the pickup and delivery problem with time windows. In *Proceedings 13th IEEE International Conference on Tools with Artificial Intelligence (ICTAI 2001)*, pp. 160–167.
23. Lin, C., Choy, K., Ho, G., Chung, S., & Lam, H. (2014). Survey of green vehicle routing problem: Past and future trends. *Expert Systems with Applications, 41*(4, Part 1), 1118–1138.
24. Mangiaracina, R., Perego, A., Salvadori, G., & Tumino, A. (2017). A comprehensive view of intelligent transport systems for urban smart mobility. *International Journal of Logistics Research and Applications, 20*(1), 39–52.
25. Mendes, R. S., Wanner, E. F., Martins, F. V. C., & Sarubbi, J. F. M. (2017). Dimensionality reduction approach for many-objective vehicle routing problem with demand responsive transport. In *Proceedings of Evolutionary Multi-criterion Optimization – 9th International Conference, EMO 2017*, Münster, 19–22 Mar 2017 (pp. 438–452).
26. Messaoud, E., El Bouzekri El Idrissi, A., & Alaoui, A. E. (2018). The green dynamic vehicle routing problem in sustainable transport. In *2018 4th International Conference on Logistics Operations Management (GOL)*, Apr 2018 (pp. 1–6).
27. Mondragon, A. E. C., Lalwani, C. S., Mondragon, E. S. C., Mondragon, C. E. C., & Pawar, K. S. (2012). Intelligent transport systems in multimodal logistics: A case of role and contribution through wireless vehicular networks in a sea port location. *International Journal of Production Economics, 137*(1), 165–175.
28. Mori, N., Kita, H., & Nishikawa, Y. (1996). Adaptation to a changing environment by means of the thermodynamical genetic algorithm. In *Parallel Problem Solving from Nature – PPSN IV* (Vol. 1141, pp. 513–522).
29. Ouertani, N., Ben Ramdhan, H., Krichen, S., Nouaouri, I., & Allaoui, H. (2018). A new evolutionary method to deal with the dynamic vehicle routing problem. In *2018 IEEE International Conference on Technology Management, Operations and Decisions (ICTMOD)* (pp. 1–5).
30. Pillac, V., Gendreau, M., Guéret, C., & Medaglia, A. L. (2013). A review of dynamic vehicle routing problems. *European Journal of Operational Research, 225*(1), 1–11.
31. Črepinšek, M., Liu, S. H., & Mernik, M. (2013). Exploration and exploitation in evolutionary algorithms: A survey. *ACM Computing Surveys, 45*, Article 35. https://doi.org/10.1145/2480741.2480752.
32. Sedighizadeh, D., & Mazaheripour, H. (2018). Optimization of multi objective vehicle routing problem using a new hybrid algorithm based on particle swarm optimization and artificial bee colony algorithm considering precedence constraints. *Alexandria Engineering Journal, 57*(4), 2225–2239.
33. Solomon, M. M. (1987). Algorithms for the vehicle routing and scheduling problems with time window constraints. *Operations Research, 35*(2), 254–265.
34. Stanciu, E. A., Moise, I. M., & Nemtoi, L. M. (2012). Optimization of urban road traffic in intelligent transport systems. In *2012 International Conference on Applied and Theoretical Electricity (ICATE)*, Oct 2012 (pp. 1–4).
35. Tan, K. C., Lee, L. H., Zhu, K. Q., & Ou, K. (2001). Heuristic methods for vehicle routing problem with time windows. *AI in Engineering, 15*(3), 281–295.
36. Toth, P., & Vigo, D. (2001). The vehicle routing problem. In *An overview of vehicle routing problems* (pp. 1–26). Philadelphia: Society for Industrial and Applied Mathematics.
37. Ursem, R. K. (2002). Diversity-guided evolutionary algorithms. In *Proceedings of the 7th International Conference on Parallel Problem Solving from Nature*, PPSN VII (pp. 462–474).
38. Wang, X., & Cao, H. (2008). A dynamic vehicle routing problem with backhaul and time window. In *2008 IEEE International Conference on Service Operations and Logistics, and Informatics*, Oct 2008 (Vol. 1, pp. 1256–1261).

39. Yang, Z., van Osta, J., Veen, B. D. V., van Krevelen, R., van Klaveren, R., Stam, A., Kok, J. N., Bäck, T., & Emmerich, M. (2017). Dynamic vehicle routing with time windows in theory and practice. *Natural Computing, 16*(1), 119–134.
40. Yao, Z., Wang, Y., Xiao, W., Zhao, B., & Peng, B. (2019). A two-level rolling optimization model for real-time adaptive signal control. *Algorithms, 12*(2), 38.
41. Zografos, K. G., & Androutsopoulos, K. N. (2004). A heuristic algorithm for solving hazardous materials distribution problems. *European Journal of Operational Research, 152*(2), 507–519.

Index

M. M. Banat, S. Paiva (eds.), *Smart Technologies for Smart Cities*,
EAI/Springer Innovations in Communication and Computing,
https://doi.org/10.1007/978-3-030-39986-3

The manufacturer's authorised representative in the EU is Springer Nature Customer Service Centre GmbH, Europaplatz 3, 69115 Heidelberg, Germany. If you have any concerns regarding our products, please contact ProductSafety@springernature.com

Printed and bound by CPI Group (UK) Ltd, Croydon, CR0 4YY
30/09/2026
02225648-0004